AF094742

Preparation and Processing Technology of Advanced Magnesium Alloys

Preparation and Processing Technology of Advanced Magnesium Alloys

Editor

Ruizhi Wu

Basel • Beijing • Wuhan • Barcelona • Belgrade • Novi Sad • Cluj • Manchester

Editor
Ruizhi Wu
Harbin Engineering University
Harbin
China

Editorial Office
MDPI AG
Grosspeteranlage 5
4052 Basel, Switzerland

This is a reprint of articles from the Special Issue published online in the open access journal *Metals* (ISSN 2075-4701) (available at: https://www.mdpi.com/journal/metals/special_issues/198X794AE1).

For citation purposes, cite each article independently as indicated on the article page online and as indicated below:

Lastname, A.A.; Lastname, B.B. Article Title. *Journal Name* **Year**, *Volume Number*, Page Range.

ISBN 978-3-7258-1699-6 (Hbk)
ISBN 978-3-7258-1700-9 (PDF)
doi.org/10.3390/books978-3-7258-1700-9

© 2024 by the authors. Articles in this book are Open Access and distributed under the Creative Commons Attribution (CC BY) license. The book as a whole is distributed by MDPI under the terms and conditions of the Creative Commons Attribution-NonCommercial-NoDerivs (CC BY-NC-ND) license.

Contents

Marie Moses, Madlen Ullmann and Ulrich Prahl
Influence of Aluminum Content on the Microstructure, Mechanical Properties, and Hot Deformation Behavior of Mg-Al-Zn Alloys
Reprinted from: *Metals* 2023, 13, 1599, doi:10.3390/met13091599 1

Anastasia A. Akhmadieva, Anton P. Khrustalev, Mikhail V. Grigoriev, Ilya A. Zhukov and Alexander B. Vorozhtsov
Structure, Phase Composition, and Mechanical Properties of ZK51A Alloy with AlN Nanoparticles after Heat Treatment
Reprinted from: *Metals* 2024, 14, 71, doi:10.3390/met14010071 24

Madlen Ullmann, Kristina Kittner and Ulrich Prahl
Cold Formability of Twin-Roll Cast, Rolled and Annealed Mg Strips
Reprinted from: *Metals* 2024, 14, 121, doi:10.3390/met14010121 41

Qing Ji, Xiaochun Ma, Ruizhi Wu, Siyuan Jin, Jinghuai Zhang and Legan Hou
BCC-Based Mg–Li Alloy with Nano-Precipitated $MgZn_2$ Phase Prepared by Multidirectional Cryogenic Rolling
Reprinted from: *Metals* 2022, 12, 2114, doi:10.3390/met12122114 51

Qing Miao, Lantao Zhu, Wenke Wang, Zhihao Wang, Bin Shao, Wenzhen Chen, et al.
Evolution of the Microstructure and Mechanical Properties of AZ31 Magnesium Alloy Sheets during Multi-Pass Lowered-Temperature Rolling
Reprinted from: *Metals* 2022, 12, 1811, doi:10.3390/met12111811 63

Yingjie Li, Hui Yu, Chao Liu, Yu Liu, Wei Yu, Yuling Xu, et al.
High Strain Rate Deformation Behavior of Gradient Rolling AZ31 Alloys
Reprinted from: *Metals* 2024, 14, 788, doi:10.3390/met14070788 77

Bingchun Jiang, Zejun Wen, Peiwen Wang, Xinting Huang, Xin Yang, Minghua Yuan and Jianjun Xi
Study on the Optimization of the Preparation Process of ZM5 Magnesium Alloy Micro-Arc Oxidation Hard Ceramic Coatings and Coatings Properties
Reprinted from: *Metals* 2024, 14, 594, doi:10.3390/met14050594 92

Junchi Liu, Hang Yin, Zhengyi Xu, Yawei Shao and Yanqiu Wang
Improving the Corrosion Resistance of Micro-Arc Oxidization Film on AZ91D Mg Alloy through Silanization
Reprinted from: *Metals* 2024, 14, 569, doi:10.3390/met14050569 104

Ying Wang, Weichen Xu, Xiutong Wang, Quantong Jiang, Yantao Li, Yanliang Huang and Lihui Yang
Research on Dynamic Marine Atmospheric Corrosion Behavior of AZ31 Magnesium Alloy
Reprinted from: *Metals* 2022, 12, 1886, doi:10.3390/met12111886 118

Xinhe Yang, Yang Jin, Ruizhi Wu, Jiahao Wang, Dan Wang, Xiaochun Ma, et al.
Simultaneous Improvement of Strength, Ductility and Damping Capacity of Single β-Phase Mg–Li–Al–Zn Alloys
Reprinted from: *Metals* 2023, 13, 159, doi:10.3390/met13010159 134

Article

Influence of Aluminum Content on the Microstructure, Mechanical Properties, and Hot Deformation Behavior of Mg-Al-Zn Alloys

Marie Moses *, Madlen Ullmann and Ulrich Prahl

Institute of Metal Forming, Technische Universität Bergakademie Freiberg, Bernhard-von-Cotta-Straße 4, 09599 Freiberg, Germany; madlen.ullmann@imf.tu-freiberg.de (M.U.); ulrich.prahl@imf.tu-freiberg.de (U.P.)
* Correspondence: marie_moses@outlook.de

Abstract: This study compares AZ91 with AZ31 to investigate the influence of a higher Al content on the resulting microstructure, mechanical properties, and hot deformation behavior. While AZ31 exhibits a globular structure after casting, AZ91 shows a fully developed dendritic structure due to the promotion of dendrites. A heat treatment helped to homogenize AZ31, dissolved a large part of the Mg-Al precipitations in AZ91, and formed globular grains in AZ91. Due to the impact of Al on constitutional supercooling, AZ91 exhibits smaller grains than AZ31. Because of the strengthening of the solid solution, AZ91 also exhibits higher strength and hardness compared to AZ31. Cylindric compression tests of the heat-treated samples were conducted at different temperatures (300–400 °C) and strain rates ($0.1 \times 10 \text{ s}^{-1}$). The main dynamic recrystallization (DRX) mechanisms in AZ31 and AZ91 are twinning-induced DRX and discontinuous DRX. It was detected that $Mg_{17}Al_{12}$ precipitates at the grain boundaries in AZ91, which influences the grain size through pinning. Similar results could be conducted in rolling trials. Although both alloys have similar grain sizes after rolling, AZ91 exhibits higher strengths, while AZ31 shows higher ductility. This can be explained by the solid solution strengthening in AZ91 and less brittle $Mg_{17}Al_{12}$ precipitations in AZ31.

Keywords: AZ31; AZ91; microstructure; mechanical properties; flow curve; magnesium alloy; lightweight

Citation: Moses, M.; Ullmann, M.; Prahl, U. Influence of Aluminum Content on the Microstructure, Mechanical Properties, and Hot Deformation Behavior of Mg-Al-Zn Alloys. *Metals* **2023**, *13*, 1599. https://doi.org/10.3390/met13091599

Academic Editor: Ruizhi Wu

Received: 19 August 2023
Revised: 9 September 2023
Accepted: 13 September 2023
Published: 15 September 2023

Copyright: © 2023 by the authors. Licensee MDPI, Basel, Switzerland. This article is an open access article distributed under the terms and conditions of the Creative Commons Attribution (CC BY) license (https://creativecommons.org/licenses/by/4.0/).

1. Introduction

The use of magnesium alloys in the automotive industry is particularly interesting because of their lower density compared to other metallic materials and high specific strength. There are various other fields of application for magnesium alloys: aerospace, automotive, medical, electronic, sports, and defense technology [1–3]. Magnesium alloys can be divided into two categories: cast and wrought alloys. AZ91 is a very common example of a cast magnesium alloy, while AZ31 is a very popular wrought magnesium alloy. Both alloys contain 9 resp. 3 wt.% aluminum, and 1 wt.% zinc. Aluminum plays a key role in both magnesium alloys.

A higher Al content helps to improve the castability due to a larger solidification interval (compare 180 K for AZ91 and 130 K for AZ31 [4]). During solidification, the primary Mg matrix (α-Mg) solidifies first in the temperature range of 650–600 °C. Later, the eutectic reaction (Mg-$Mg_{17}Al_{12}$) occurs below 437 °C at the grain boundaries [5]. Under equilibrium cooling conditions, the eutectic phase $Mg_{17}Al_{12}$ is expected to appear in Mg-Al alloys with an Al content of approximately 13 wt.%. But, already, Mg alloys containing more than 2 wt.% Al show the secondary phase $Mg_{17}Al_{12}$ (eutectic phase) for non-equilibrium cooling conditions. The resulting size, shape, and distribution of the secondary phase have an impact on the ductility and creep resistance of the alloy. It can form in different morphologies depending on the solidification rate and the Al resp. Zn content [6]: fully or partially divorced eutectic [7]. Aluminum also influences the grain structure that forms.

Dahle et al. describe that a low Al content determines a globular structure after casting, but a higher Al content promotes the formation of a dendritic structure after casting [5]. During solidification, Al also has an impact on constitutional supercooling. This affects the growth restriction rate and, therefore, the impact on grain refinement [8]. By suppressing columnar grains and promoting narrower columnar grains with increasing Al content in pure Mg, Al influences the size of the resulting grains [8]. This means that the addition of Al can be used as a grain refinement method.

The Al content also influences the choice of heat treatment. For the AZ31 alloy, temperatures of 400 °C and holding times of approximately 12 h are generally sufficient to dissolve the precipitates of $Mg_{17}Al_{12}$ [9,10]. Their amount is reduced as a result of the lower Al content in the alloy, and their characteristic is usually fine and easier to dissolve. In the case of the AZ91 alloy, elevated temperatures (415 °C) and longer holding times (24 h) are recommended to dissolve the secondary phase [11–13]. The fully developed dendritic structure of the AZ91 alloy transforms into a globular structure after heat treatment [13]. It might be of interest to investigate whether the grain refinement effect is still visible after heat treatment in an AZ91 alloy compared to AZ31, though longer holding times and higher temperatures are recommended for the heat treatment to dissolve the higher amount of $Mg_{17}Al_{12}$ precipitations.

Although magnesium may tend to recover due to its high stacking fault energy, dynamic recrystallization (DRX) prevails due to the few slip systems available [14]. The nucleation of recrystallized grains occurs preferentially at existing high-angle grain boundaries. Furthermore, dynamic recrystallization can be initiated in twins, deformation inhomogeneities, such as deformation and shear bands, precipitations, or particles, as well as areas close to grain boundaries [15,16]. After a strong hardening, a new microstructure can be formed without the migration of a large-angle grain boundary; this process is understood as in situ or continuous dynamic recrystallization (CDRX). According to this, CDRX resembles a recovery process in which dislocations are consumed by small-angle grain boundaries, leading to the formation of new grains with high-angle grain boundaries [17]. The discontinuous dynamic recrystallization (DDRX) is characterized by the migration of a high-angle grain boundary. The form of a necklace structure is typical here [18]. In addition, grain boundary serration and bulging are an indication for DDRX [19]. Twin-induced dynamic recrystallization (TDRX) is often found in coarse-grained material, such as homogenized material [20,21]. Additionally, TDRX also occurs predominantly at lower degrees of deformation. It is reported that twinning is suppressed with an increasing Al content in Mg-Al-Zn alloys [22,23]. The bulging mechanism is characterized by the local movement of grain boundaries and occurs only in magnesium alloys at elevated temperatures and when the dislocation arrangement is favorable [20]. The particle-stimulated nucleation (PSN) is common in magnesium alloys containing rare-earth elements and leads to weaker texture formation, as the particles, which act as nucleation sites, form grains with a more random orientation [24]. Depending on the size, distribution, and proportion of the particles, they can support or suppress recrystallization [25]. The so-called pinning effect of $Mg_{17}Al_{12}$ precipitates was also described in Mg-Al-Zn alloys with an increased Al content. The size and number of precipitates are hereby decisive in terms of the influence on the resulting dynamically recrystallized grain size and the degree of dynamic recrystallization [21,26].

Not only can the Al content affect the DRX mechanism during hot deformation, but also a finer grain size before hot deformation [16,27]. In contrast, other publications show that the initial microstructure does not play a role [28]. Therefore, it is of particular interest to investigate the influence of grain refinement during solidification and heat treatment on hot deformation behavior. This study attempts to induce a finer grain size in AZ91 than in AZ31 after casting through the influence of the alloying element Al. This research is part of broader research exploring grain refinement methods and their effect and inheritance before, during, and after deformation. The AZ31 alloy is compared with the AZ91 in terms of its deformation behavior, with the initial condition before deformation (heat-treated state) being appropriately matched to the alloy (different temperatures and holding times due to the amount of Al). Further, this study investigates whether a variation in alloying content influences the microstructure, texture, and mechanical properties during annealing and hot deformation.

2. Materials and Methods

The AZ31 and AZ91 ingots were melted and cast at 750 °C, resp., 720 °C (different temperatures due to the solidification interval) in dies with a diameter of 16 mm and a length of 250 mm. The steel dies were smoothened with boron nitride and graphite and preheated to 450 °C. Table 1 shows the chemical composition of the cast samples compared to the standard. AZ91 shows a slightly higher Al content compared to the standard.

Table 1. Chemical composition of the AZ31 and AZ91 samples compared to the standard in wt.%.

Alloy	Al	Zn	Mn	Balance	Mg
AZ31	2.8	0.7	0.3	0.0	96.2
AZ31 standard (DIN EN 12438) [29]	2.5–3.5	0.6–1.4	0.2–1.0	-	-
AZ91	9.6	0.6	0.2	0.0	89.6
AZ91 standard (ASTM B93) [30]	8.5–9.5	0.45–0.9	0.17–0.4	-	-

Heat treatment was carried out in an air radiation furnace. The following heat treatments were chosen according to the literature: 400 °C, 12 h for AZ31 [9,10], and 415 °C, 24 h for AZ91 [11–13]. After heat treatment, the samples were rapidly cooled in water.

Cylindric compression samples (10 mm in diameter, 18 mm in height) were milled from these heat-treated samples. Hot compression tests were conducted in the Warmumformsimulator (WUMSI) at temperatures of 300, 350, and 400 °C and strain rates of 0.1, 1, and 10 s^{-1} to an equivalent logarithmic strain of 1. After heating the samples in an air radiation furnace for 15 min at elevated temperatures, they were deformed and then water quenched. Graphite was used as a lubricant. The recorded flow curves were corrected for temperature and friction.

Groove rolling tests were performed on a three-high-standing rolling mill at the Institute of Metal Forming at TU Bergakademie Freiberg. Oval rolling samples with a height of 9.2 mm and a width of 20 mm were milled from the cast and heat-treated samples. Before rolling, the samples were heated to an elevated temperature (350 °C for AZ31 and 300 °C for AZ91 due to cracking during rolling at higher temperatures) in an air radiation furnace. Within five rolling steps, using a square-oval calibration, the rolling samples were deformed to a final diameter of 9.8 mm. The rolling speed was 1.5 m/s and the summed logarithmic strain reached 0.65.

Samples of the AZ31 and AZ91 alloys in the cast, heat-treated, compressed, and rolled state were embedded in the longitudinal section, ground with SiC abrasive paper, and polished with OP-chem and MD-Chem OPS 300 polishing cloths. The cast and heat-treated samples were etched using picric acid (5 mL glacial acetic acid, 6 g picric acid, 10 mL distilled water, and 100 mL ethanol) for 10–20 s. The deformed samples were etched using nitric acid (20 mL glacial acetic acid, 1 mL nitric acid, 20 mL distilled water, and 60 mL ethanol) for 45–50 s. Optical characterization was performed with the digital microscope VHX-6000 (Keyence Corporation, Osaka, Japan). The grain sizes were measured using the linear intercept method.

Furthermore, the scanning electron microscope Jeol JSM 7800 F (Tokyo, Japan) was used to take SE pictures and measure alloy composition using energy-dispersive X-ray spectroscopy (EDX). In addition, texture analysis was carried out by electron backscattering diffraction (EBSD) using an accelerating voltage of 20 kV and a step size depending on grain size (0.1–2 µm). The EBSD data were analyzed using the MTEX MATLAB toolbox [31].

Quasi-static tensile tests of the heat-treated and rolled samples were conducted at the AG100 at room temperature. The testing speed reached 0.625 mm/min. The sample form was B (according to standard DIN EN 50125) [32] with a measurement diameter of

5 mm and a measurement length of 25 mm. Vickers HV 10 hardness measurements were carried out on cast, heat-treated, and rolled samples using the ZHU250 (Zwick/Roell, Ulm, Germany).

3. Results and Discussion
3.1. As-Cast and Heat-Treated State

Figure 1 shows the microstructure of the cast (a) AZ31 and (b) AZ91 magnesium alloy. A globular structure can be seen for the AZ31, while the AZ91 alloy exhibits a fully developed dendritic structure. This can be mainly explained by the higher Al content, which promotes the formation of dendrite arms [5].

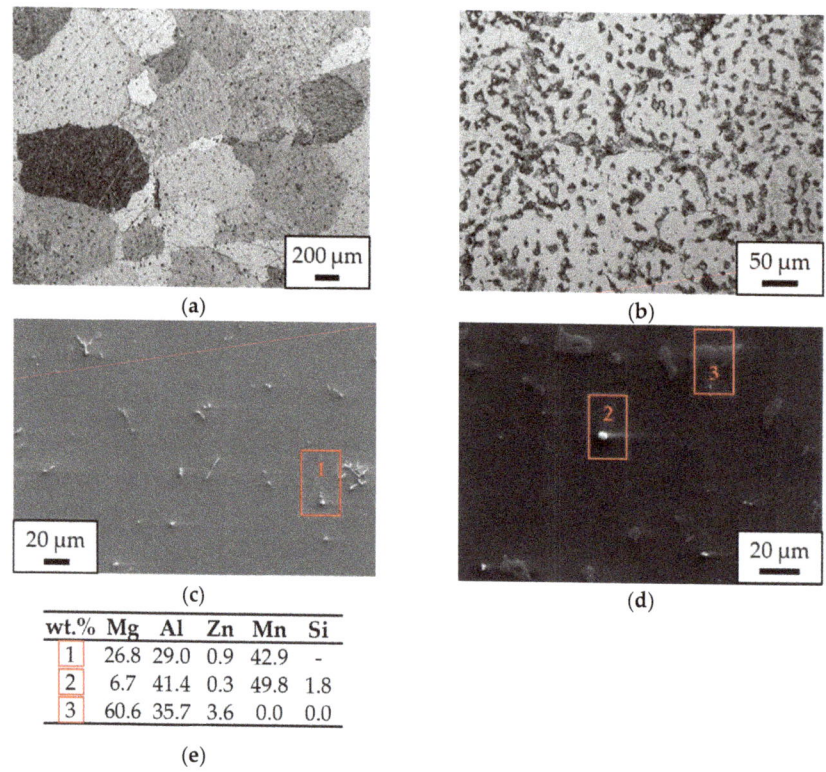

Figure 1. Microstructure of the cast magnesium alloy (**a**) AZ31 and (**b**) AZ91. SEM images of the cast magnesium alloy (**c**) AZ31 and (**d**) AZ91 with EDX measurements in (**e**).

As mentioned before, Mg-Al alloys that do not solidify under equilibria conditions are expected to form the secondary eutectic phase $Mg_{17}Al_{12}$ when having more than 2 wt.% Al [5]. Although this phase could not be clearly detected in AZ31, the precipitate could be measured in AZ91 (see Figure 1d, point (3)). Furthermore, the Al_8Mn_5 precipitates are probably visible in both alloys (see Figure 1c,d, points (1) and (2)) [33]. Due to the higher Al content, AZ91 exhibits a higher amount of the secondary eutectic phase, which develops between the dendrite arms. The precipitations are evenly distributed. While the Al_8Mn_5 precipitates have a rather round, sometimes rod-like shape, the $Mg_{17}Al_{12}$ phase is extensive and significantly longer and larger. The size of an Al-Mn-based precipitate is approximately 2–5 µm and that of the Mg-Al precipitation is approximately 5–15 µm. The Al content in the magnesium matrix of AZ91 is also higher compared to AZ31 (4 wt.% compared to 2 wt.%).

Due to heat treatment, the grain structure of AZ91 transforms into a globular grain structure (see Figure 2). Furthermore, the SEM images reveal that the $Mg_{17}Al_{12}$ precipitates could be mainly dissolved in the matrix. This is supported by the fact that the Al content in the magnesium matrix of the heat-treated AZ91 sample is higher than in the cast AZ91 sample (9 wt.% compared to 4 wt.%). The Al content of the AZ31 matrix is also increasing (4 wt.% compared to 2 wt.%). The Al_8Mn_5 precipitates, measured at points one and two in Figure 2 [33], are still visible in both alloys after heat treatment. They cannot dissolve [34], and their size remains the same.

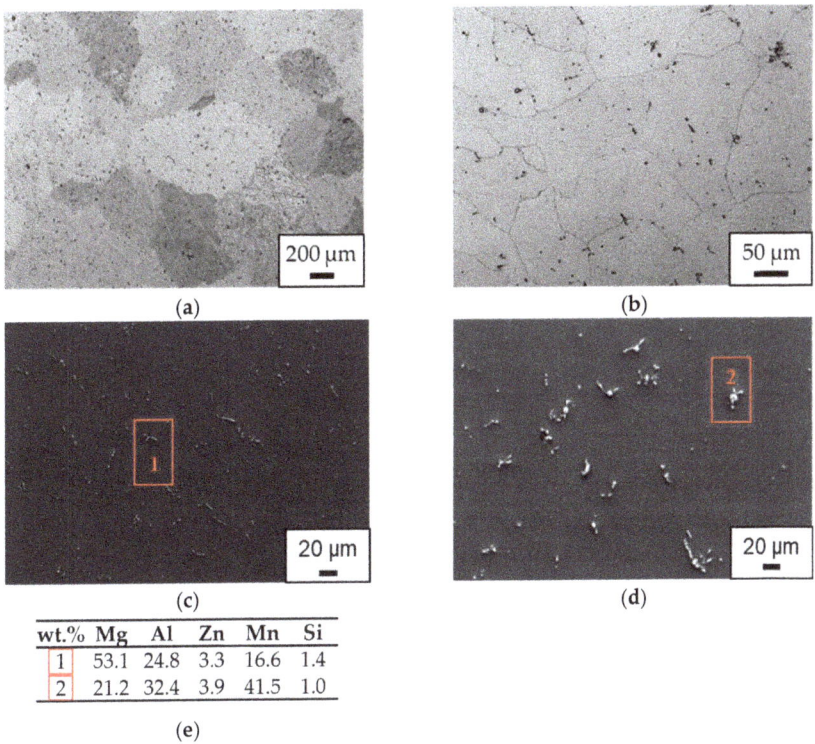

Figure 2. Microstructure of the heat-treated magnesium alloy (**a**) AZ31 (400 °C, 12 h) and (**b**) AZ91 (415 °C, 24 h). SEM images of the heat-treated magnesium alloy (**c**) AZ31 and (**d**) AZ91 with EDX measurements in (**e**).

Finally, the EBSD maps of the cast and heat-treated samples were examined. Figure 3a,b shows whether AZ31 or AZ91 offers a preferred orientation because no color (red, blue, green) is represented in priority. This is also supported by the inverse pole figures shown in Figure 3c,d for AZ31 and AZ91, respectively. No clear orientation is present whether in the radial casting direction or longitudinal casting direction.

Figure 4 shows the grain size distribution of the cast and heat-treated AZ31 and AZ91. In general, the grain size is coarse because the material is poured and heat treated, but AZ91 exhibits in both, as-cast and heat-treated state, smaller grains than AZ31. This might be primarily due to the grain refining effect of the higher Al content and its impact on constitutional supercooling (in the case of the cast state). In the case of the heat-treated samples, a lower grain size of AZ91 could be due to inheritance effects, though a higher temperature and longer holding time were used to dissolve $Mg_{17}Al_{12}$ precipitations.

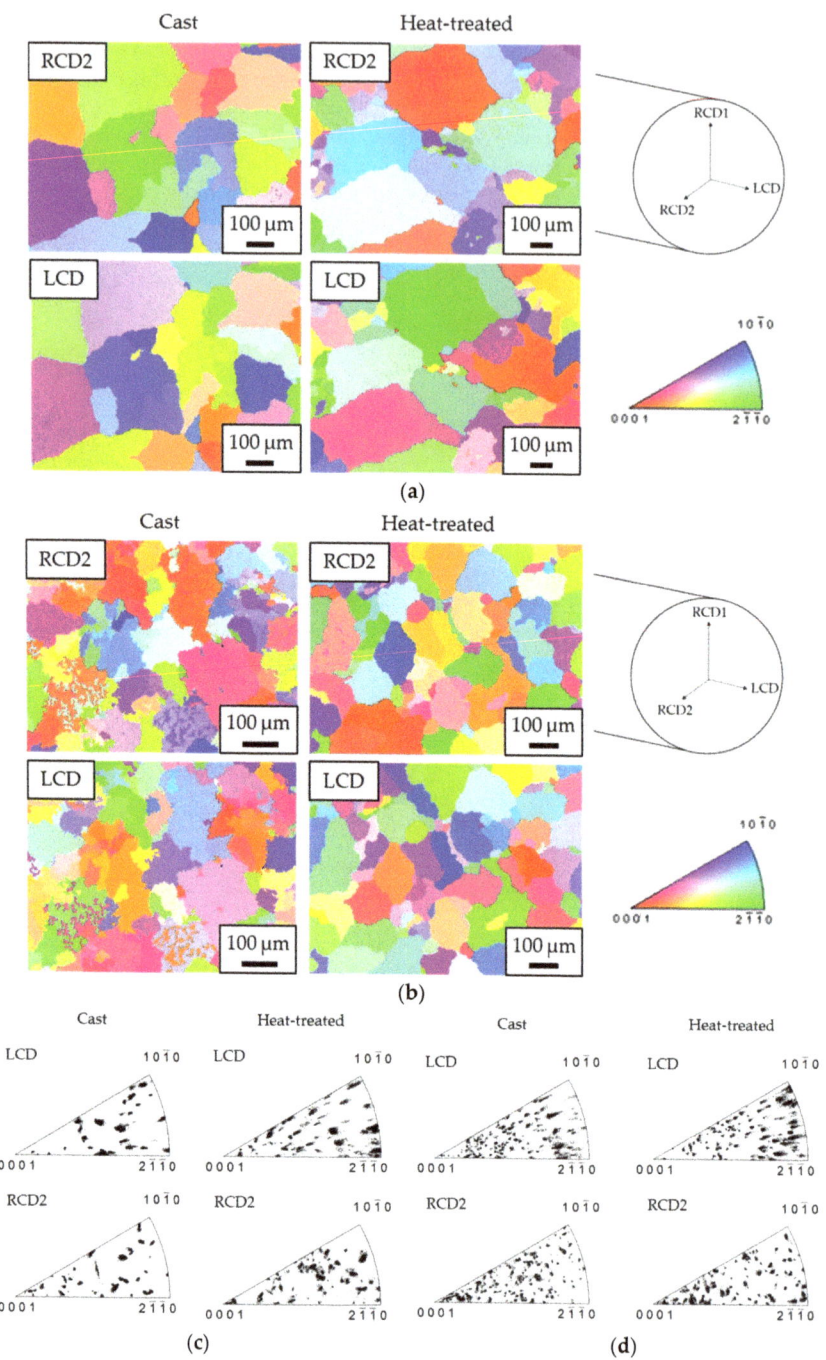

Figure 3. EBSD of the cast and heat-treated magnesium alloy (**a**) AZ31 and (**b**) AZ91 in radial casting direction 2 (RCD2) and longitudinal casting direction (LCD). Inverse pole figures of the cast and heat-treated magnesium alloy (**c**) AZ31 and (**d**) AZ91 in RCD2 and LCD.

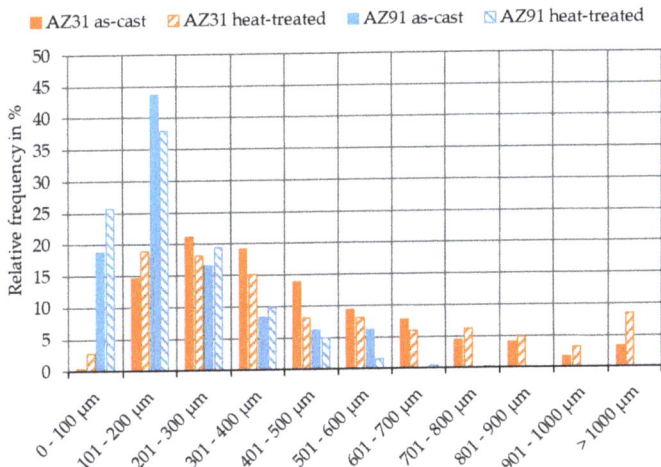

Figure 4. Grain size distribution of the heat-treated magnesium alloy AZ31 and AZ91 compared to the cast state.

The hardness values in Table 2, measured at room temperature (RT), show that the hardness of AZ91 significantly decreases after heat treatment. This is attributed to the dissolution of Al in the Mg matrix. In contrast to that, no clear trend could be determined in the case of AZ31, as the hardness drop is in the measurement error. AZ91 exhibits higher hardness values due to the higher amount of Al and its solid solution strengthening, as well as the grain refinement effect. It is already reported in the literature that the hardness increases with the Al content [35].

Table 2. Hardness values of the magnesium alloys AZ31 and AZ91 in cast and heat-treated state.

Hardness HV10 (RT)	As Cast	Heat-Treated
AZ31	49 ± 5	46 ± 2
AZ91	74 ± 2	61 ± 2

Table 3 shows the mechanical properties of the heat-treated alloys AZ31 and AZ91 conducted at room temperature. While AZ31 exhibits higher ductility, AZ91 shows higher strengths in the heat-treated state. This is mainly due to the higher Al content in AZ91 and the effect of Al on solid solution strengthening and grain size. In contrast to that, due to the lower Al content, AZ31 exhibits a higher stacking fault energy (SFE) [36], which might contribute to a higher ductility of the material.

Table 3. Mechanical properties of the alloys AZ31 and AZ91 in the heat-treated state at room temperature.

Mechanical Properties (RT)	Yield Strength (MPa)	Tensile Strength (MPa)	Elongation at Break (%)
AZ31	63 ± 7	192 ± 16	11 ± 1
AZ91	86 ± 4	229 ± 11	8 ± 1

The solidification rate of AZ91 in permanent mold casting is typically 10 K/s [37]. By examining the secondary dendrite arm spacing (*SDAS*) and calculating the solidification rate, a comparison between the investigation and the literature can be drawn. An *SDAS* of 33 μm was measured for AZ91. Using the equation

$$SDAS = 10.5 \, t_f^{0.4}, \tag{1}$$

a local solidification time t_f of 17.5 s was calculated. Assuming a solidification interval T_s of 163 K for AZ91 [38], a solidification rate of 9.3 K/s was then calculated using the formula:

$$Solidification\ rate = \frac{\Delta T_s}{t_f}. \qquad (2)$$

This shows that the solidification rate of the AZ91 alloy is a typical solidification rate for a material in permanent mold casting. The solidification rate of AZ31 could not be identified metallographically due to the globular structure.

However, the solidification rates of AZ31 and AZ91 during casting were additionally measured using a thermocouple placed in the middle of a steel die (see Figure 5). The solidification curves show a similar course though casting was carried out at different temperatures. The bend in the curves may be attributed to the solidification of the material (transition of liquid + Mg to Mg in the phase diagram). Due to the difference in the Al content, the change in the curve takes place at a lower temperature for AZ91 and affects the solidification rate strongly. However, the longer the time, the more the solidification rate of AZ31 is expected to be similar to that of AZ91.

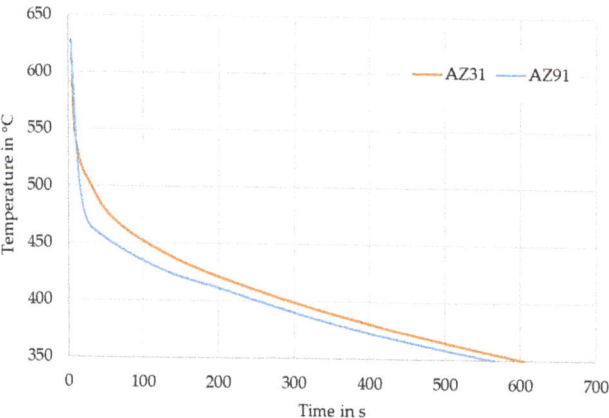

Figure 5. The solidification rate of AZ31 and AZ91.

All in all, a finer grain size was induced in the AZ91 alloy compared to the AZ31 alloy after casting due to the solute element Al and its influence on constitutional supercooling. The finer grain size is still present after heat treatment. A fully developed dendritic structure in the case of AZ91 after solidification could be formed into a globular structure after heat treatment. In addition, $Mg_{17}Al_{12}$ precipitations were dissolved during heat treatment, which resulted in a lower hardness after heat treatment for AZ91. In the case of the AZ31 alloy, a globular structure was already present after casting. No preferred orientation is present in both alloys. Due to solid solution strengthening, AZ91 exhibits higher strengths, while AZ31 shows a higher ductility in the heat-treated state.

3.2. Compressed State

Figure 6 shows the flow curves of the hot compressed alloys AZ31 and AZ91 after casting and heat treatment. The flow curves of AZ31 and AZ91 are characteristic flow curves for these magnesium alloys and correspond to the literature [14,39,40]. In the diagrams, a representative curve from each of the multiple trials is shown. The scatter of the curves is quite low with 5 MPa. As both alloys show flow curves with an increase and decrease in flow stress with increasing logarithmic strain, it is assumed that dynamic recrystallization is more dominant than dynamic recovery.

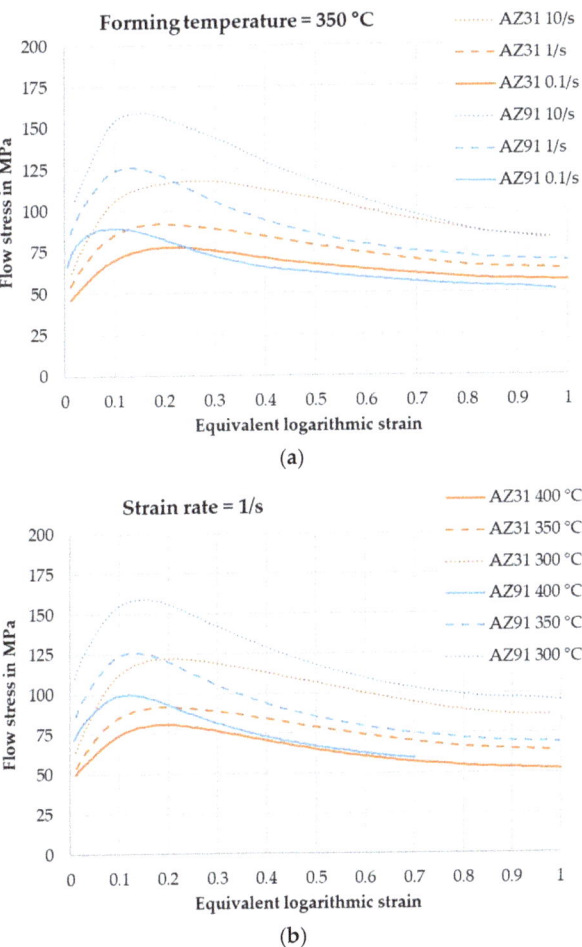

Figure 6. Flow curve of the AZ31 and AZ91 alloy at (**a**) a forming temperature of 350 °C and at (**b**) a strain rate of 1/s.

AZ91 often exhibits a steeper slope and decline than AZ31 at the same forming temperature and strain rate. It was already reported in the literature that Mg-Al alloys with a higher Al content tend to show higher peak flow stresses compared to Mg-Al alloys with a lower Al content due to the strengthening effect of the solute [35]. In addition, AZ91 exhibited a finer grain size before hot deformation, contributing to hardening. The maximum flow stresses of the AZ91 magnesium alloy are shifted to lower logarithmic strains and higher flow stresses compared to AZ31 at the same forming temperature and strain rate. This was also already reported in the literature for Mg-Al alloys with a higher Al content [19]. The flatter flow curves of AZ31 compared to AZ91 could be an indication that AZ31 has a higher tendency toward CDRX relative to AZ91, which might have a higher tendency toward DDRX. This may be attributed to a higher stacking fault energy (SFE) of AZ31 relative to AZ91 due to its lower Al content [19,36,40].

Figure 7 depicts the microstructure of the compressed AZ31 and AZ91 samples at a forming temperature of 350 °C, a strain rate of 1 s^{-1}, and a logarithmic strain of about 0.2. The black dots in the microstructure of AZ91 are probably Mg-Al precipitations that are precipitating during hot deformation. The SEM picture of the AZ91 alloy also reveals that the secondary phase precipitates (see Figure 7c,d) along the grain boundaries and inside the grains. The precipitation of $Mg_{17}Al_{12}$ was not observed for AZ31. In both alloys, probably Al_8Mn_5 particles could be detected (an example of AZ91 is shown; see EDX measurement in Figure 7d). It is reported in the literature that Mg-Al-Zn alloys with higher Al content (e.g., AZ80) tend to show the initiation of DDRX at lower strains than Mg-Al-Zn alloys with a lower Al content (e.g., AZ31) [19]. Therefore, the AZ91 alloy is expected to show a higher amount of DRX grains compared to AZ31 at the logarithmic strain of 0.2. This can be observed in the microstructural pictures in Figure 7a,b. The newly formed grains are aligned along the original grain boundaries, indicating DDRX. In addition, the lower initial grain size of the AZ91 alloy probably contributes to a higher amount of DRX grains due to its effects on enhancing the nucleation rate. It was reported earlier that the initial microstructure influences the DRX proportion in an AZ31 alloy [41,42].

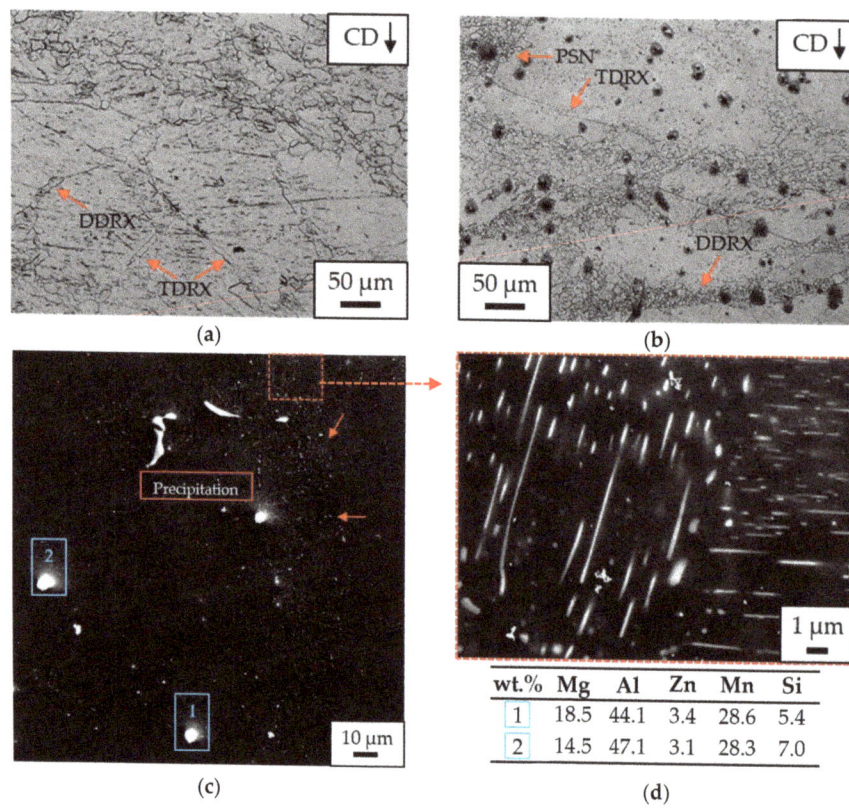

Figure 7. Microstructure of the hot-deformed magnesium alloy (**a**) AZ31 and (**b**) AZ91 at a forming temperature of 350 °C, a strain rate of 1 s^{-1}, and a logarithmic strain of approximately 0.2. (**c**) SEM image of the AZ91 alloy showing the precipitation of $Mg_{17}Al_{12}$. (**d**) The magnification of precipitation inside the grains and the result of the EDX measurement shown in (**c**).

The mean DRX grain size is 10 ± 4 μm for AZ31 and 9 ± 3 μm for AZ91 at a forming temperature of 350 °C and a strain rate of 1 s^{-1}. It is reported that the higher the strain rate, the more comparable the DRX grain size will be between lower (AZ31) and higher (AZ80) Al-alloyed Mg-Al-Zn alloys [19]. This might explain the similar DRX grain size of both alloys.

As the secondary phase in AZ91 also precipitates along the grain boundaries, recrystallized grains and $Mg_{17}Al_{12}$ are often found together at the grain boundaries. Interestingly, the grain size near the precipitation is even smaller than farther away from the precipitation. Therefore, the precipitations might act as nucleation points (PSN) and have a pinning effect on the recrystallized grains. Depending on their size, the precipitations can either promote (coarse particles) or hinder (fine dispersoids) recrystallization as they can act as nucleation sites or suppress the growth of DRX-grains [26]. Regarding the size of the precipitations in Figure 7, it is assumed that they might act as nucleation points for DRX.

In addition, twinning is observed as a deformation mechanism in both alloys. Due to a limited number of slip systems and an activation of twinning at a lower or equivalent stress, twinning is quite common in Mg alloys. Nevertheless, twinning is also influenced by the Al content. It is reported that the formation of twin lamellar structures is suppressed in Mg-Al-Zn alloys by increasing the Al content [22]. While double twins were more dominant in AZ31, extension twins without a lamellae structure nucleated in AZ91 [22]. Figure 8 presents the misorientation angle distribution of the magnesium alloys AZ31 and AZ91. The peak between 85 and 90 degrees may indicate $\{10\bar{1}2\}$ extension twins, while the peak at 35 degrees indicates $\{10\bar{1}1\}$-$\{10\bar{1}2\}$ double twins. These double twins are generated due to $\{10\bar{1}2\}$ re-twinning inside a $\{10\bar{1}1\}$ compression twin and occur more frequently under compression when the stress direction is along the c-axis [43]. It seems to be that double twins are more dominant in AZ31 compared to AZ91, though both types are present in both alloys. It is known that the formation of $\{10\bar{1}2\}$ extension twins is much easier due to lower critical resolved shear stresses compared to $\{10\bar{1}1\}$ compression twins [23]. Nevertheless, the formation of $\{10\bar{1}1\}$ compression twins will be triggered due to basal texture formation under compression [21]. Then, $\{10\bar{1}1\}$-$\{10\bar{1}2\}$ double twins will be formed to reduce the strain caused by $\{10\bar{1}1\}$ compression twins [21]. Further, it was reported that the presence of $Mg_{17}Al_{12}$ precipitations may lead to a lesser extent of twinning in AZ91 compared to AZ31 [23]. During hot deformation, $Mg_{17}Al_{12}$ precipitates as shown in Figure 7c,d. Depending on the size of the precipitation, they can act as barriers toward twin boundary migration and hinder the twin propagation rate [23].

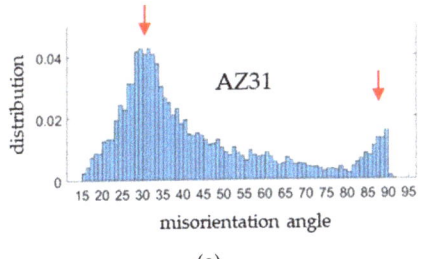
(a)

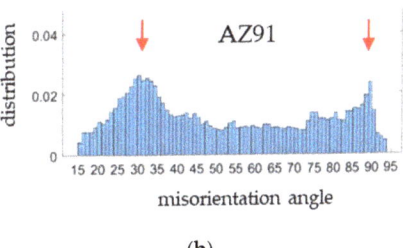
(b)

Figure 8. Misorientation angle distribution of (**a**) AZ31 and (**b**) AZ91 magnesium alloy at forming temperature of 350 °C and at a strain rate of 1/s.

In addition to DDRX, PSN, and TDRX, it is interesting to discuss the mechanism of CDRX during the deformation of AZ31 and AZ91. It was reported that the precipitation of $Mg_{17}Al_{12}$ in an AZ91 alloy hinders dislocation movement and, therefore, leads to dislocation pile-up near the precipitations and inside the grains, resulting in a promotion of the CDRX process [21]. This might be visible in the following EBSD picture of the AZ91 alloy during hot compression (see Figure 9b,d). For AZ31, it was reported that non-basal slip promotes CDRX in general, but DDRX is more dominant at higher temperatures compared to CDRX [18,39,44]. As the deformation took place at 350 °C, it is assumed that DDRX might be more dominant than CDRX in AZ31.

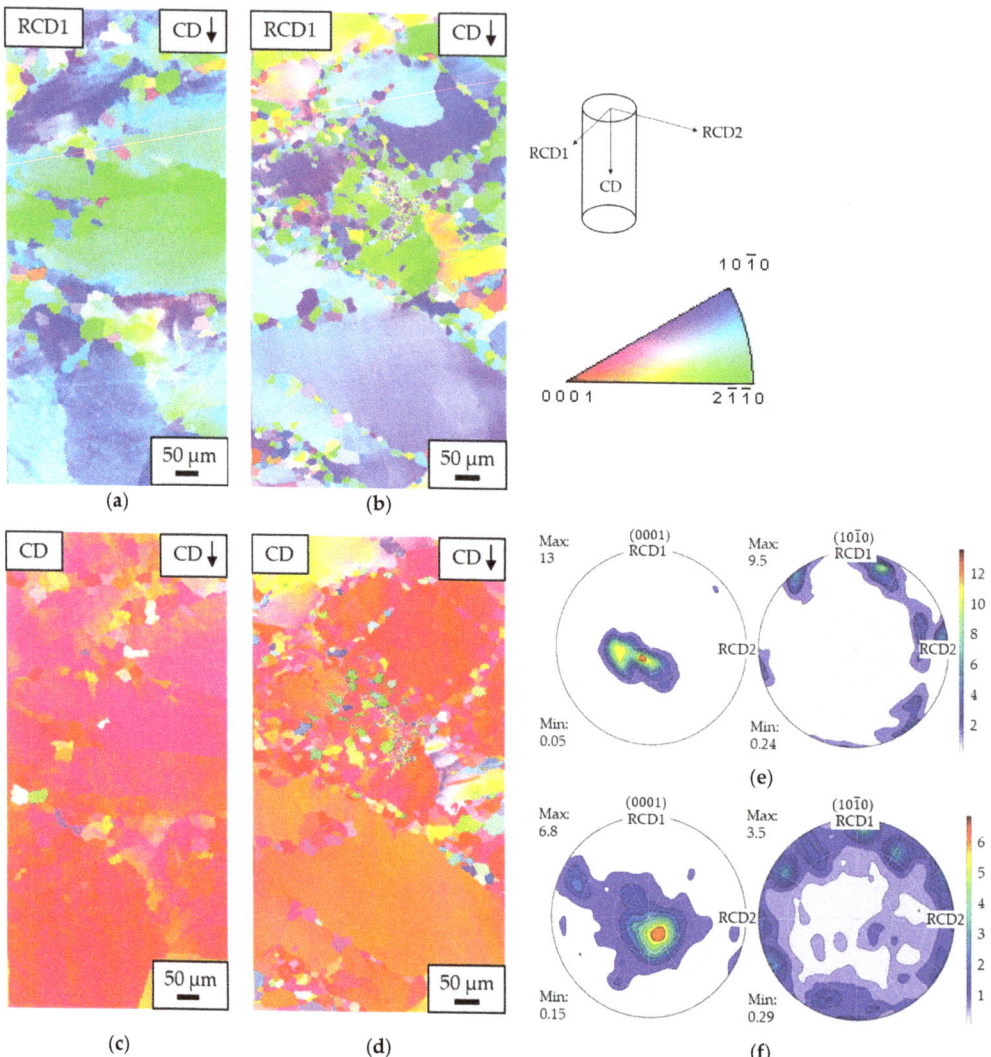

Figure 9. EBSD maps of the magnesium alloy AZ31 transverse to compression direction (CD) (**a**) in radial compression direction 1 (RCD1), (**c**) in CD and AZ91 transverse to CD (**b**) in RCD1, (**d**) in CD at a forming temperature of 350 °C, a strain rate of 1 s^{-1}, and a logarithmic strain of about 0.2. Pole figures showing (**e**) AZ31 and (**f**) AZ91 magnesium alloy.

The EBSD maps in Figure 9 show that a basal texture is present due to hot deformation. Since the grains in Figure 9a,b are primarily colored green or blue, this indicates that the hexagonal unit cells are aligned along the compression direction (CD). Therefore, the basal planes (marked red) are visible from the CD (see Figure 9c,d). The pole figures in Figure 9e,f also reveal that a basal texture is present. The intensity of the texture is lower in AZ91 compared to AZ31, probably due to the higher amount of DRX grains shown on the map, which promotes a less basal texture. It is reported in the literature that dynamically recrystallized grains in AZ91 contribute to a weaker texture, while the addition of Al limits the activation of prismatic slip, resulting in a strong texture for non-recrystallized grains [22].

To summarize this section, it can be stated that flow curves were obtained from the hot compression tests for AZ31 and AZ91. Both flow curves showed hardening and softening, a typical indication of dynamical recrystallization. By observing the microstructures at a forming temperature of 350 °C, a strain rate of 1 s^{-1}, and a logarithmic strain of about 0.2, it was found that Mg$_{17}$Al$_{12}$ precipitates at the grain boundaries in AZ91 and acts as nucleation sites for DRX. In both alloys, twinning was observed, acting as an initiation for DRX. In addition, DDRX is assumed to be dominant compared to CDRX in both alloys. A basal texture is present in both alloys after compression at a forming temperature of 350 °C, a strain rate of 1 s^{-1}, and a logarithmic strain of about 0.2. The texture intensity in AZ91 is lower, which was ascribed to more DRX grains in AZ91 with a random orientation. As the start of DRX is probably shifted to lower logarithmic strains due to the higher Al content, it is expected to see a higher amount of DRX grains in AZ91. In addition, the lower initial grain size in AZ91 may contribute to a higher amount of DRX grains.

3.3. Hot Deformation Behavior

To describe the dynamic recrystallization processes, the flow stress maxima, which depend on the strain rate and temperature, were obtained from the warm flow curves. This forms the basis for determining the activation energy Q for dynamic recrystallization using the logarithmic Arrhenius equation (see Equation (3)). This represents thermal activation during hot forming. Furthermore, the average model coefficients A (material constant), α (fitting parameter), and n (hardening exponent) can be graphically determined by using the slopes in the following diagrams and showing their relationships (see Figure 10). Furthermore, the Zener–Hollomon parameter Z, which summarizes the influence of the forming rate and thermal activation during hot forming, can be calculated using the formula:

$$Z = \dot{\varphi} e^{(\frac{Q}{RT})} = A[\sinh(\alpha\, \sigma_{max})]^n. \quad (3)$$

$\dot{\varphi}$ represents the effective strain rate in s^{-1}, T the thermodynamic temperature, R the ideal gas constant (8.314 J/(mol·K)), and σ_{max} the peak stresses. The following formulas result for AZ31 and AZ91:

$$\text{AZ31}\ (r^2 = 0.99): Z = \dot{\varphi} e^{(\frac{140,120}{RT})} = 1.535 \cdot 10^{11}[\sinh(0.010\, \sigma_{max})]^{7.965} \quad (4)$$

$$\text{AZ91}\ (r^2 = 0.99): Z = \dot{\varphi} e^{(\frac{146,780}{RT})} = 5.638 \cdot 10^{11}[\sinh(0.008\, \sigma_{max})]^{7.128}. \quad (5)$$

The activation energies between 105 and 185 kJ/mol are common for AZ alloys during hot deformation [45]. The calculated activation energy for AZ31 is 140 kJ/mol and, for AZ91, it is 147 kJ/mol. However, higher-alloyed AZ alloys show a higher activation energy for hot deformation compared to low-alloy AZ alloys [46,47]. In this present case, AZ91 exhibits only a slightly higher value than AZ31. All the values are above the self-diffusion value for magnesium (135 kJ/mol) [15].

Regarding the hardening parameter n, in the literature, it is said that there is a creep of dislocation climb present $n > 5$ [48,49]. Therefore, it is concluded that the dislocation climb creep is the dominant mechanism for both alloys since AZ31 and AZ91 exhibit higher values than 5 (7.965, resp., 7.128).

With the help of the so-called Kocks–Mecking plot, the development of the grain structure during deformation can be described. To evaluate the flow behavior, the strengthening rate (calculated slopes from the flow curves) was plotted against the flow stress. First, a strong and linear drop in hardening can be seen, which turns into an increase shortly before the maximum flow stress, which differs from the linear increase. This critical stress marks the beginning of dynamic recrystallization. A critical logarithmic strain for recrystallization can be assigned to that stress. Figure 11a shows the logarithmic strain for recrystallization in dependence on the Zener–Hollomon parameter, which has been determined mathematically. The critical degree decreases with increasing forming temperature (see Figure 11b) and reduced strain rate, so the start of dynamic recrystallization shifts to lower logarithmic strains. The influence of forming temperature might be greater than the influence of the strain rate. AZ91 exhibits significantly lower logarithmic strains in dependence on Z and the forming temperature compared to AZ31.

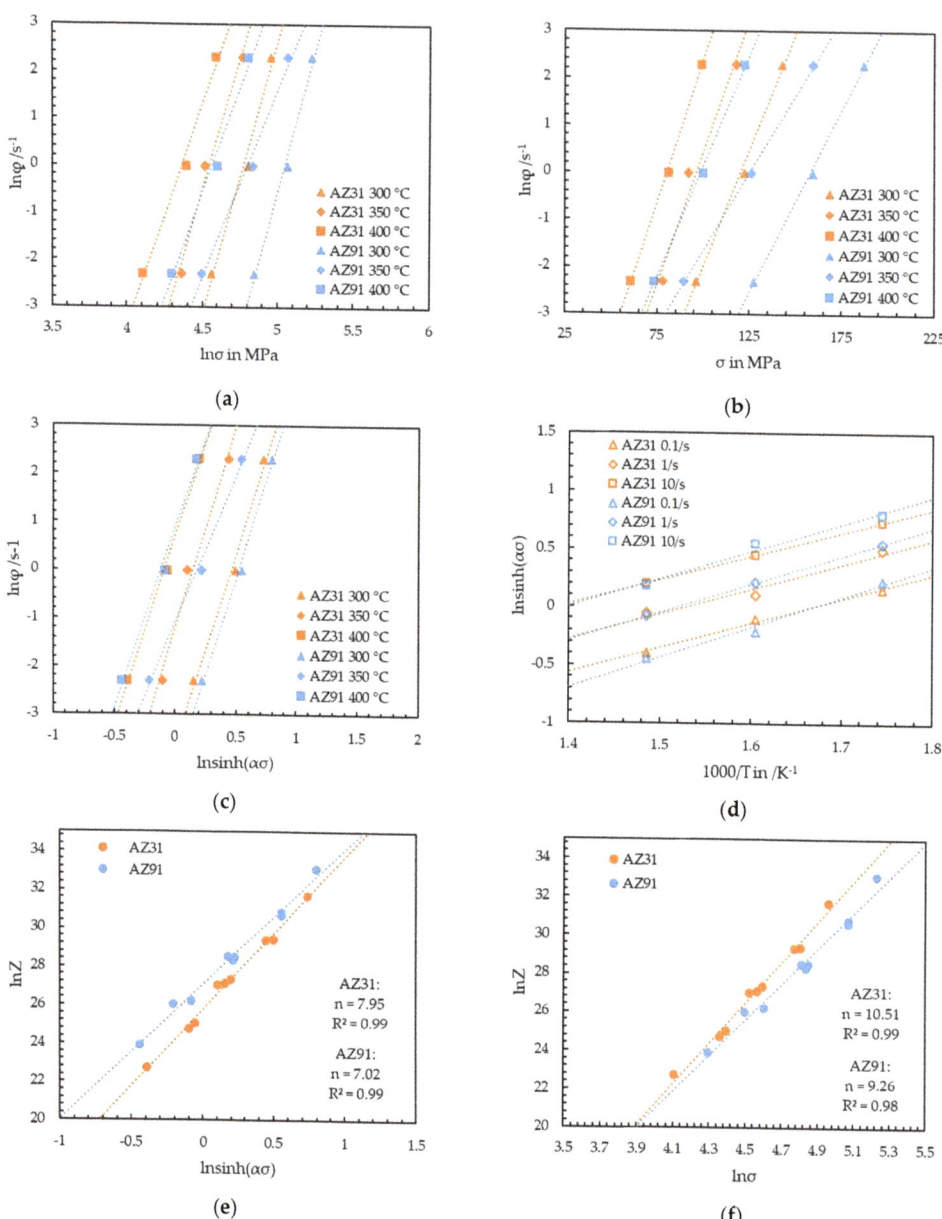

Figure 10. Relationship between $\ln\dot{\varphi}$ and (**a**) $\ln(\sigma)$, (**b**) σ, and (**c**) $\ln\sinh(\alpha\sigma)$, the relationship between (**d**) $\ln\sinh(\alpha\sigma)$ and $1000/T$, and the relationship between $\ln(Z)$ and (**e**) $\ln\sinh(\alpha\sigma)$ and (**f**) $\ln(\sigma)$ for AZ31 and AZ91 (cast, heat-treated, and hot-deformed).

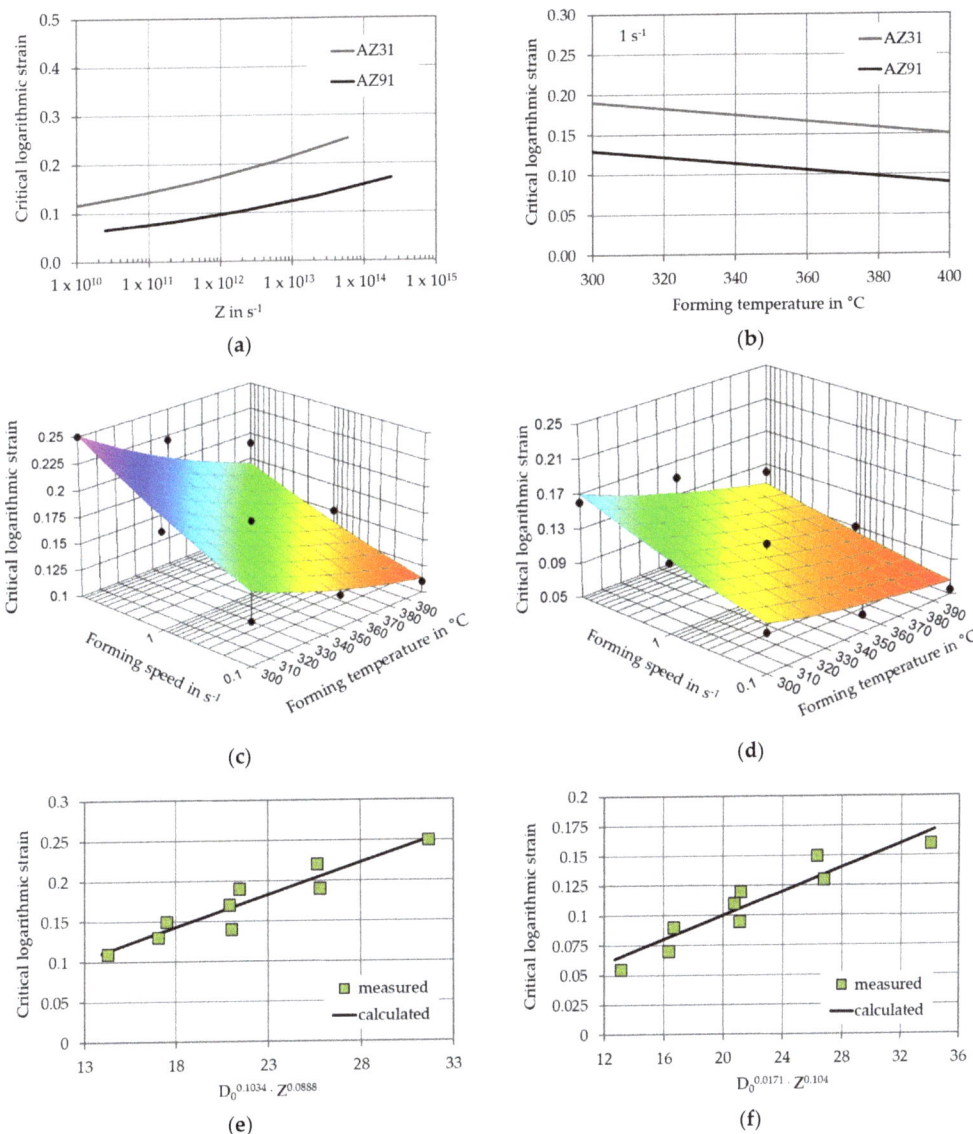

Figure 11. (**a**) Critical logarithmic strain in dependence on Zener–Hollomon parameter, (**b**) critical logarithmic strain in dependence of forming temperature for AZ31 and AZ91, (**c**) critical logarithmic strain in dependence of strain rate and temperature for AZ31, and (**d**) for AZ91, (**e**) comparison between measured and calculated results for AZ31 and (**f**) AZ91. Both alloys are cast, heat-treated, and hot-deformed.

Furthermore, the critical degree for all speeds and temperatures was graphically displayed from the experimental data and compared with the calculated results (see Figure 11c–f). For AZ31, the critical logarithmic strain for dynamic recrystallization for all speeds and temperatures lies between 0.11 and 0.25. However, AZ91 exhibits significantly lower values for the critical degree for all speeds and temperatures (0.055–0.16). Accordingly, the increased Al content contributes to the fact that the start of dynamic recrystallization is shifted to lower logarithmic strains.

The critical logarithmic strain for AZ31 and AZ91 can be summarized in the following equation:

$$\text{AZ31} \left(r^2 = 0.89\right) : \varphi_c = a_1 \cdot D_0^{a_2} \cdot Z^{a_3} = 0.0079 \cdot D_0^{0.1034} \cdot Z^{0.0888} \tag{6}$$

$$\text{AZ91} \left(r^2 = 0.89\right) : \varphi_c = a_1 \cdot D_0^{a_2} \cdot Z^{a_3} = 0.0050 \cdot D_0^{0.01715} \cdot Z^{0.1038}. \tag{7}$$

3.4. Rolled State

In the case of the AZ31 magnesium alloy, a good comparison can be made between the compression tests and the rolling tests. For the evaluation, however, it should be noted that the deformation rate in the rolling tests was significantly higher than 1 s^{-1}, as shown in Section 3.3. Therefore, in Figure 12, the EBSD maps of the first rolling pass (approximately 350 °C and about 15 s^{-1}) will be analyzed. With higher deformation rates, the start of dynamical recrystallization will be shifted to higher logarithmic strains. These are mostly achieved in the first pass so that recrystallization starts despite the increased strain rate. The discontinuous dynamical recrystallization (DDRX) mechanism could be observed after the first pass, probably due to higher strain rates [50,51]. In addition, the twinning mechanism was dominant due to the presence of a coarse initial grain structure, which is known to contribute to TDRX [20]. It is also reported that higher strain rates can contribute to twinning [19], which was the case during rolling. In total, the same mechanisms were observed that were already discussed with the compression sample at 350 °C and 1 s^{-1}.

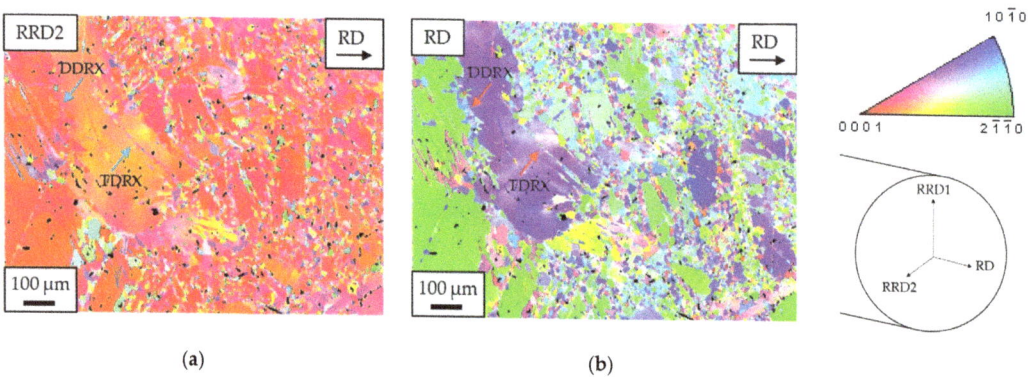

Figure 12. EBSD maps of the rolled AZ31 magnesium alloy (first pass) transverse to RD, (**a**) in RRD2, and (**b**) in RD.

For the AZ91 magnesium alloy, it should be noted that the rolling temperature was reduced to 300 °C due to cracking during rolling at 350 °C. A reduced forming temperature contributes to the fact that the critical degree for DRX is shifted to higher logarithmic strains. However, this was already exceeded in the first rolling pass. Figure 13 shows the microstructure of the first pass of the AZ91 alloy. As a coarse grain structure is present, TDRX occurs. Furthermore, DDRX could be observed. Although DDRX is prevalent at higher forming temperatures, higher strain rates also contribute to this mechanism [50,51].

Moreover, the precipitation of the secondary phase can be detected. At higher temperatures, continuous precipitation occurs and the higher the forming temperature, the greater the amount of precipitation, but the less the effect of pinning [52]. At lower temperatures, discontinuous precipitation is present, and lower temperatures also lead to less precipitation of the secondary phase, but it enhances their pinning effect [14,52]. Pinning of the newly formed recrystallized grain boundaries restricts their growth and affects a lower DRX grain size. As precipitation takes place at the grain boundaries of the original grains, pinning and particle-stimulated nucleation is present there.

The microstructure of the magnesium alloys AZ31 and AZ91 after five passes of hot rolling is shown in Figure 14. AZ91 exhibits a lower grain size, which may be due to the lower rolling temperature used for AZ91. Additionally, the secondary phase $Mg_{17}Al_{12}$ precipitates at the grain boundaries during hot deformation (see Figure 14d). An EDX analysis reveals that the Al content in the matrix of AZ91 is slightly decreasing due to precipitation (8 wt.% compared to 9 wt.%). This was not observed for AZ31, as the Al content stays the same (4 wt.%). However, Al_8Mn_5 particles are present in both alloys.

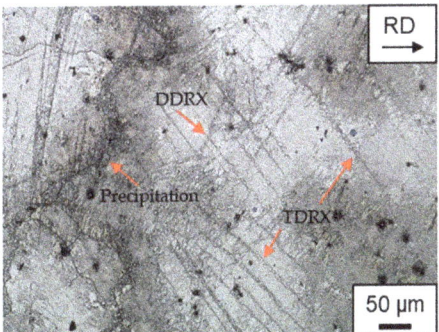

Figure 13. Microstructure of the AZ91 magnesium alloy after the first rolling pass, showing TDRX, DDRX, and precipitation.

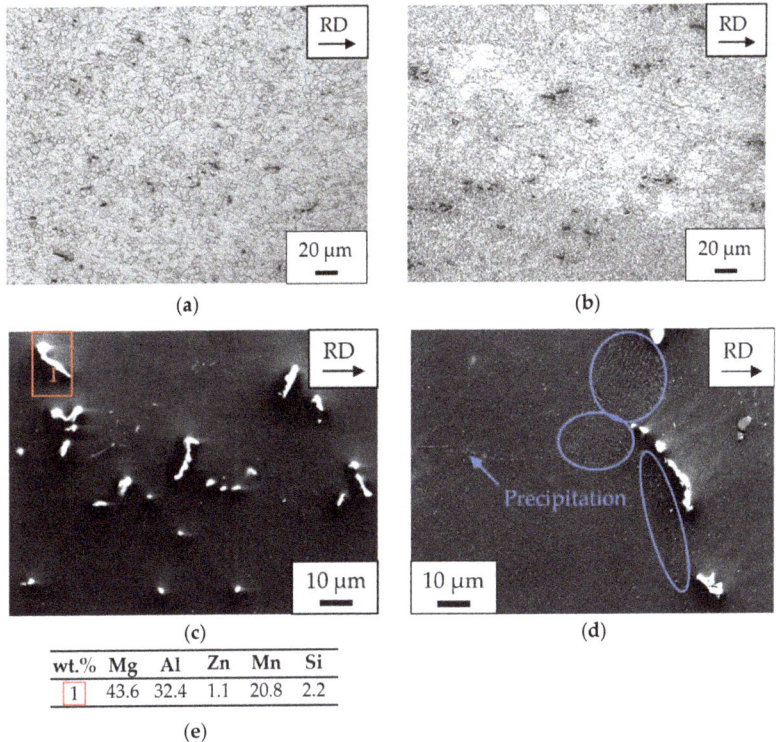

Figure 14. Microstructure of the rolled (five passes) magnesium alloys (**a**) AZ31 and (**b**) AZ91. SEM images of the rolled (five passes) magnesium alloys (**c**) AZ31 and (**d**) AZ91; blue marked is precipitation of $Mg_{17}Al_{12}$. (**e**) EDX measurement. RD is the rolling direction.

Compared to the heat-treated state, the grain size of AZ31 and AZ91 could be significantly decreased during hot rolling. After five passes, the mean grain size of AZ31 is 5 ± 2 μm, while AZ91 has a mean grain size of 4 ± 1 μm.

The EBSD maps for both alloys are shown in Figure 15. A wire-rolling texture is present in which the hexagonal unit cells are horizontal and aligned transversely to the wire axis. The pole figures represent the intensity of the texture. AZ91 exhibits a lower texture intensity than AZ31. On the one hand, it may be because a lower rolling temperature was used for AZ91 and after the fifth pass there were still many randomly oriented, newly formed grains. On the other hand, the secondary phase precipitates during hot rolling, which promotes the formation of randomly oriented grains [17].

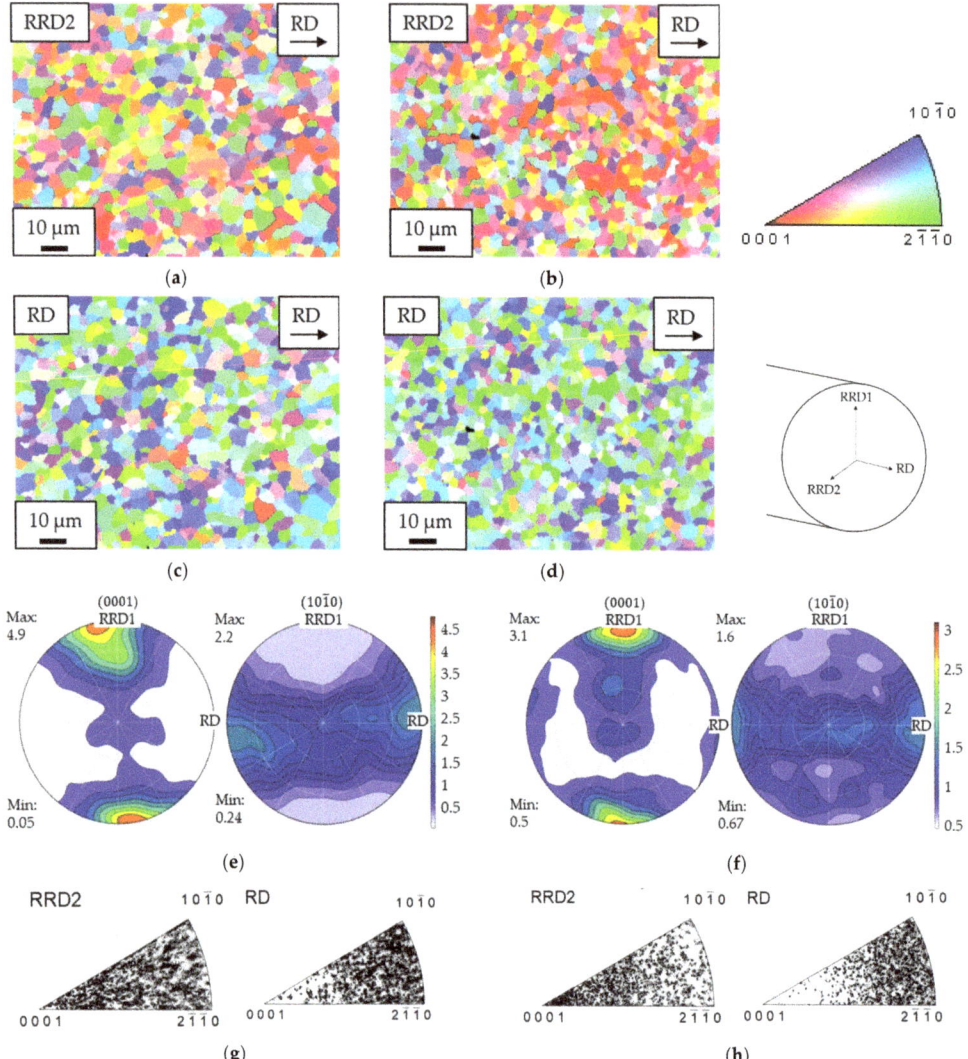

Figure 15. EBSD maps of the rolled magnesium alloy AZ31 transverse to rolling direction (RD) (**a**) in radial rolling direction 2 (RRD2), (**c**) in RD, and AZ91 transverse to RD (**b**) in RRD2 and (**d**) in RD. Pole figures of the rolled magnesium alloys (**e**) AZ31 and (**f**) AZ91. Inverse pole figures of (**g**) AZ31 and (**h**) AZ91 magnesium alloy.

Table 4 shows the hardness values of AZ31 and AZ91 in the heat-treated and rolled state. After hot rolling, a significant hardness increase is present for both alloys. On the one hand, the hardness increase is attributed to grain refinement because of dynamical recrystallization with increasing rolling pass, see Figure 16. On the other hand, the Al content (solid solution strengthening) influences hardness, as the values of the AZ91 alloy were significantly higher than the values of the AZ31 magnesium alloy.

Table 4. Hardness values of the magnesium alloys AZ31 and AZ91 in the heat-treated and rolled state.

Hardness HV10	Heat-Treated	Rolled
AZ31	46 ± 2	68 ± 2
AZ91	61 ± 2	92 ± 1

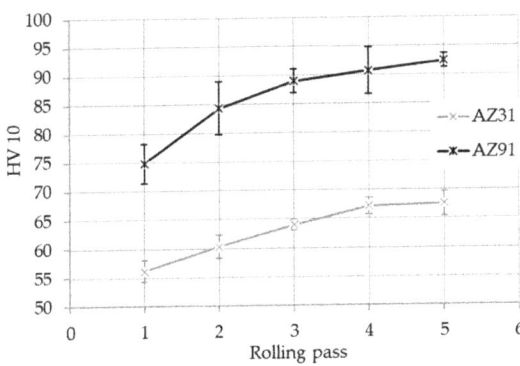

Figure 16. Hardness increases of AZ31 and AZ91 during rolling.

Table 5 shows the mechanical properties of the alloys AZ31 and AZ91 after five passes of hot rolling. An increase in strength and ductility is present for both alloys compared to the heat-treated state (cf. Table 3). Furthermore, the standard deviations could be reduced in the case of yield and tensile strength. This might be due to a reduction in grain size and refinement of the material. AZ31 still exhibits improved ductility, while AZ91 shows higher strengths. This is attributed to the lower, respectively, higher Al content, which leads to a higher SFE in AZ31 and, resp., solid solution strengthening in AZ91 [35,36].

Table 5. Mechanical properties of the alloys AZ31 and AZ91 in the rolled state.

Mechanical Properties	Yield Strength (MPa)	Tensile Strength (Mpa)	Elongation at Break (%)
AZ31	208 ± 3	283 ± 6	14 ± 0
AZ91	310 ± 1	384 ± 7	9 ± 2

Within the rolled state, the same DRX mechanisms were observed in both alloys as described in Section 3.2. Though the initial grain size differed, nearly the same mean grain size was obtained in both alloys after five passes of rolling. AZ31 and AZ91 exhibited a rolling texture after five passes of rolling. However, the texture intensity was lower for AZ91. The mechanical properties, as well as the hardness, could be improved by wire rolling. As in the heat-treated state, AZ31 exhibited a higher ductility while AZ91 exhibited higher strengths.

After five passes of hot rolling, which corresponds to a large strain, and a corresponding microstructure formation during DRX, the inheritance effect of a fine initial grain size is no longer evident, as it was in the heat-treated state. An initial grain size influences how the DRX is initiated. The increased Al content, which influenced a fine initial grain size in AZ91 compared to AZ31, not only affects the initiation of DRX but also contributes to the final properties via solid solution strengthening and precipitation.

4. Conclusions

This study showed that a higher Al content causes a larger constitutional supercooling in the magnesium melt, which leads to a finer grain size in AZ91 compared to AZ31. Then, it investigated whether the alloying element influences the texture, DRX mechanisms, and mechanical properties during annealing and hot deformation. This study can be summarized as follows:

- In the cast state, the globular structure of the AZ31 alloy differs from the fully developed dendritic microstructure of the AZ91 alloy. This can be explained by the higher Al content, which promotes the formation of dendrites. After heat treatment, a globular grain structure is visible for both alloys. The grain size of the AZ91 alloy is lower compared to AZ31 after casting due to the higher Al content and its impact on constitutional supercooling. No preferred orientation is present for both alloys in the cast or heat-treated state. The hardness is increased in AZ91 compared to AZ31 due to the higher amount of Al and the solid solution strengthening effect. At room temperature, AZ91 exhibits higher strength due to the solid solution strengthening of the higher Al content, while AZ31 shows a higher elongation at the break due to its higher SFE.
- The solidification rate of both alloys shows a similar course but is also affected by the Al content. The calculated solidification rate of AZ91 was about 9.3 K/s, which is comparable to a typical solidification rate in the literature.
- The warm flow curves of the AZ91 alloy exhibit a steeper slope and decline compared to the AZ31 alloy. This is mainly due to the strengthening effect of the solute. Further, the maximum flow stresses of the AZ91 alloy are shifted to lower logarithmic strains and higher flow stresses compared to the AZ31 alloy. According to the flow curves, dynamic recrystallization (DRX) is probably more dominant than dynamic recovery. The calculated activation energy for DRX was 140 kJ/mol for AZ31 and 147 kJ/mol for AZ91. The start of the DRX for AZ91 is shifted to lower logarithmic strains compared to AZ31. It is, therefore, expected to see a higher amount of DRX grains in AZ91. The main DRX mechanisms are TDRX and DDRX in both alloys. It was detected that $Mg_{17}Al_{12}$ precipitates at the grain boundaries in AZ91, which influences the grain size through pinning (PSN).
- The same results on the DRX mechanisms were obtained in rolling tests, where a higher deformation rate was present. After five passes of rolling, a fine DRX grain size is present in both alloys. The wire rolling texture intensity in AZ91 is lower compared to AZ31. This is attributed to more randomly oriented grains that are present in AZ91 due to the precipitation of the secondary phase. Within five passes of rolling, the hardness can be increased in both alloys. Although AZ31 exhibits higher ductility in the rolled state, AZ91 shows higher strength values. This is probably due to the lower, respectively, higher Al content, which leads to a higher SFE in AZ31 and, resp., solid solution strengthening in AZ91.
- Due to a microstructure formation during DRX, the inheritance effect of a fine initial grain size is no longer evident, as it was in the heat-treated state. The increased Al content in AZ91 compared to AZ31 influences not only a fine grain size in the initial state but also the initiation of the DRX and final properties via solid solution strengthening and the precipitation of the secondary phase.

Author Contributions: Conceptualization, M.M.; methodology, M.M.; formal analysis, M.M.; investigation, M.M.; data curation, M.M.; writing—original draft preparation, M.M.; writing—review and editing, M.U. and U.P.; supervision, M.U. and U.P.; project administration, M.U. All authors have read and agreed to the published version of the manuscript.

Funding: This research received no external funding.

Data Availability Statement: The data presented in this study are available on request from the corresponding author. The data are not publicly available due to ongoing research.

Acknowledgments: We would like to acknowledge Jonas Lachmann's support regarding the SEM investigations (Institute of Material Science, TU Bergakademie Freiberg).

Conflicts of Interest: The authors declare no conflict of interest.

References

1. Siengchin, S. A review on lightweight materials for defence applications: Present and future developments. *Def. Technol.* **2023**, *24*, 1–17. [CrossRef]
2. Jayasathyakawin, S.; Ravichandran, M.; Baskar, N.; Anand Chairman, C.; Balasundaram, R. Mechanical properties and applications of Magnesium alloy—Review. *Mater. Today Proc.* **2020**, *27*, 909–913. [CrossRef]
3. Elambharathi, B.; Kumar, S.D.; Dhanoop, V.U.; Dinakar, S.; Rajumar, S.; Sharma, S.; Kumar, V.; Li, C.; Eldin, E.M.T.; Wojciechowski, S. Novel insights on different treatment of magnesium alloys: A critical review. *Heliyon* **2022**, *8*, e11712. [CrossRef]
4. Schichtel, G. *Magnesium Taschenbuch*; VEB Verlag Technik: Berlin, Germany, 1954.
5. Dahle, A.K.; Lee, Y.C.; Nave, M.D.; Schaffer, P.L.; StJohn, D.H. Development of the as-cast microstructure on magnesium–aluminium alloys. *J. Light Met.* **2001**, *1*, 61–72. [CrossRef]
6. Długosz, P.; Bochniak, W.; Ostachowski, P.; Molak, R.; Duarte Guigou, M.; Hebda, M. The Influence of Conventional or KOBO Extrusion Process on the Properties of AZ91 (MgAl9Zn1) Alloy. *Materials* **2021**, *14*, 6543. [CrossRef]
7. Kang, J.-H.; Park, J.; Song, K.; Oh, C.-S.; Shchyglo, O.; Steinbach, I. Microstructure analyses and phase-field simulation of partially divorced eutectic solidification in hypoeutectic Mg-Al Alloys. *J. Magnes. Alloys* **2022**, *10*, 1672–1679. [CrossRef]
8. Lee, Y.C.; Dahle, A.K.; StJohn, D.H. The Role of Solute in Grain Refinement of Magnesium. *Metall. Mater. Trans. A* **2000**, *31*, 2895–2906. [CrossRef]
9. Shen, M.J.; Wang, X.J.; Li, C.D.; Zhang, M.F.; Hu, X.S.; Zheng, M.Y.; Wu, K. Effect of submicron size SiC particles on microstructure and mechanical properties of AZ31B magnesium matrix composites. *Mater. Des.* **2014**, *54*, 436–442. [CrossRef]
10. Wang, X.; Liu, W.; Hu, X.; Wu, K. Microstructural modification and strength enhancement by SiC nanoparticles in AZ31 magnesium alloy during hot rolling. *Mater. Sci. Eng. A* **2018**, *715*, 49–61. [CrossRef]
11. Xu, Q.; Li, Y.; Ding, H.; Ma, A.; Jiang, J.; Chen, G.; Chen, Y. Microstructure and mechanical properties of SiCp/AZ91 composites processed by a combined processing method of equal channel angular pressing and rolling. *J. Mater. Res. Technol.* **2021**, *15*, 5244–5251. [CrossRef]
12. You, Z.; Jiang, A.; Duan, Z.; Qiao, G.; Gao, J.; Guo, L. Effect of heat treatment on microstructure and properties of semi-solid squeeze casting AZ91D. *China Foundry* **2020**, *17*, 219–226. [CrossRef]
13. Lee, D.H.; Lee, G.M.; Park, S.H. Difference in extrusion temperature dependences of microstructure and mechanical properties between extruded AZ61 and AZ91 alloys. *J. Magnes. Alloys* **2023**, *11*, 1683–1696. [CrossRef]
14. Ebrahimi, G.R.; Maldar, A.R.; Ebrahimi, R.; Davoodi, A. Effect of thermomechanical parameters on dynamically recrystallized grain size of AZ91 magnesium alloy. *J. Alloys Compd.* **2011**, *509*, 2703–2708. [CrossRef]
15. Gottstein, G. *Physikalische Grundlagen der Materialkunde*, 3rd ed.; Springer: Berlin/Heidelberg, Germany, 2007; ISBN 978-3-540-71104-9.
16. Mirzadeh, H. Grain refinement of magnesium alloys by dynamic recrystallization (DRX): A review. *J. Mater. Res. Technol.* **2023**, *25*, 7050–7077. [CrossRef]
17. Ghandehari Ferdowsi, M.R.; Mazinani, M.; Ebrahimi, G.R. Effects of hot rolling and inter-stage annealing on the microstructure and texture evolution in a partially homogenized AZ91 magnesium alloy. *Mater. Sci. Eng. A* **2014**, *606*, 214–227. [CrossRef]
18. Ebrahimi, M.; Wang, Q.; Attarilar, S. A comprehensive review of magnesium-based alloys and composites processed by cyclic extrusion compression and the related techniques. *Prog. Mater. Sci.* **2023**, *131*, 101016. [CrossRef]
19. Prakash, P.; Wells, M.A.; Williams, B.W. Hot deformation of cast AZ31 and AZ80 magnesium alloys—Influence of Al content on microstructure and texture development. *J. Alloys Compd.* **2022**, *897*, 1–11. [CrossRef]
20. Sitdikov, O.; Kaibyshev, R. Dynamic Recrystallization in Pure Magnesium. *Mater. Trans.* **2001**, *42*, 1928–1937. [CrossRef]
21. Xu, S.W.; Kamado, S.; Matsumoto, N.; Honma, T.; Kojima, Y. Recrystallization mechanism of as-cast AZ91 magnesium alloy during hot compressive deformation. *Mater. Sci. Eng. A* **2009**, *527*, 52–60. [CrossRef]
22. Guo, F.; Zhang, D.; Wu, H.; Jiang, L.; Pan, F. The role of Al content on deformation behavior and related texture evolution during hot rolling of Mg-Al-Zn alloys. *J. Alloys Compd.* **2017**, *695*, 396–403. [CrossRef]
23. Tahreen, N.; Chen, D.L.; Nouri, M.; Li, D.Y. Influence of aluminum content on twinning and texture development of cast Mg–Al–Zn alloy during compression. *J. Alloys Compd.* **2015**, *623*, 15–23. [CrossRef]

24. Li, X.; Jiao, F.; Al-Samman, T.; Ghosh Chowdhury, S. Influence of second-phase precipitates on the texture evolution of Mg–Al–Zn alloys during hot deformation. *Scr. Mater.* **2012**, *66*, 159–162. [CrossRef]
25. Robson, J.D.; Henry, D.T.; Davis, B. Particle effects on recrystallization in magnesium–manganese alloys: Particle-stimulated nucleation. *Acta Mater.* **2009**, *57*, 2739–2747. [CrossRef]
26. Jin, Z.-Z.; Cheng, X.-M.; Zha, M.; Rong, J.; Zhang, H.; Wang, J.-G.; Wang, C.; Li, Z.-G.; Wang, H.-Y. Effects of Mg17Al12 second phase particles on twinning-induced recrystallization behavior in Mg−Al−Zn alloys during gradient hot rolling. *J. Mater. Sci. Technol.* **2019**, *35*, 2017–2026. [CrossRef]
27. Zhu, H.; Yu, B.; Cai, J.; Bian, J.; Zheng, L. Effect of initial microstructures on microstructures and properties of extruded AZ31 alloy. *Mater. Sci. Technol.* **2023**, *39*, 1579–1591. [CrossRef]
28. Prakash, P.; Uramowski, J.; Wells, M.A.; Williams, B.W. Influence of Initial Microstructure on the Hot Deformation Behavior of AZ80 Magnesium Alloy. *J. Mater. Eng. Perform* **2023**, *32*, 2647–2660. [CrossRef]
29. DIN EN 12438:2017; Magnesium and Magnesium Alloys—Magnesium Alloys for Casting Anodes. Beuth Verlag: Berlin, Germany, 2017.
30. ASTM B93/B93M-21; Specification for Magnesium Alloys in Ingot Form for Sand Castings, Permanent Mold Castings, and Die Castings. ASTM International: West Conshohocken, PA, USA, 2021.
31. Bachmann, F.; Hielscher, R.; Schaeben, H. Texture Analysis with MTEX—Free and Open Source Software Toolbox. *SSP* **2010**, *160*, 63–68. [CrossRef]
32. DIN EN 50125:2022; Testing of Metallic Materials - Tensile Test Pieces. Beuth Verlag: Berlin, Germany, 2022.
33. Krbetschek, C.; Berge, F.; Oswald, M.; Ullmann, M.; Kawalla, R. Microstructure investigations of inverse segregations in twin-roll cast AZ31 strips. In *Magnesium Technology*; Springer: Cham, Switzerland, 2016; pp. 369–374. [CrossRef]
34. Pawar, S.; Zhou, X.; Hashimoto, T.; Thompson, G.E.; Scamans, G.; Fan, Z. Investigation of the microstructure and the influence of iron on the formation of Al8Mn5 particles in twin roll cast AZ31 magnesium alloy. *J. Alloys Compd.* **2015**, *628*, 195–198. [CrossRef]
35. Cáceres, C.; Rovera, D. Solid solution strengthening in concentrated Mg–Al alloys. *J. Light Met.* **2001**, *1*, 151–156. [CrossRef]
36. Muzyk, M.; Pakiela, Z.; Kurzydlowski, K.J. Generalized stacking fault energy in magnesium alloys: Density functional theory calculations. *Scr. Mater.* **2012**, *66*, 219–222. [CrossRef]
37. Park, S.S.; Lee, J.G.; Lee, H.C.; Kim, N.J. Development of wrought Mg alloys via strip casting. In *Essential Readings in Magnesium Technology*; Mathaudhu, S.N., Luo, A.A., Neelameggham, N.R., Nyberg, E.A., Sillekens, W.H., Eds.; Springer International Publishing: Cham, Switzerland, 2004; pp. 233–238.
38. Luo, A.A.; Fu, P.; Zheng, X.; Peng, L.; Hu, B.; Sachdev, A.K. Microstructure and Mechanical Properties of Die Cast Magnesium-Aluminium-Tin Alloys. *Magnes. Technol.* **2013**, *2016*, 341–345.
39. Fatemi-Varzaneh, S.M.; Zarei-Hanzaki, A.; Beladi, H. Dynamic recrystallization in AZ31 magnesium alloy. *Mater. Sci. Eng. A* **2007**, *456*, 52–57. [CrossRef]
40. Kim, K.; Ji, Y.; Kim, K.; Park, M. Effect of Al Concentration on Basal Texture Formation Behavior of AZ-Series Magnesium Alloys during High-Temperature Deformation. *Materials* **2023**, *16*, 2380. [CrossRef] [PubMed]
41. Pilehva, F.; Zarei-Hanzaki, A.; Fatemi-Varzaneh, S.M. The influence of initial microstructure and temperature on the deformation behavior of AZ91 magnesium alloy. *Mater. Des.* **2012**, *42*, 411–417. [CrossRef]
42. Yang, X.Y.; Sanada, M.; Miura, H.; Sakai, T. Effect of Initial Grain Size on Deformation Behavior and Dynamic Recrystallization of Magnesium Alloy AZ31. *MSF* **2005**, *488-489*, 223–226. [CrossRef]
43. Barnett, M.R.; Keshavarz, Z.; Beer, A.G.; Ma, X. Non-Schmid behaviour during secondary twinning in a polycrystalline magnesium alloy. *Acta Mater.* **2008**, *56*, 5–15. [CrossRef]
44. Lin, B.; Zhang, H.; Meng, Y.; Wang, L.; Fan, J.; Zhang, S.; Roven, H.J. Deformation behavior, microstructure evolution, and dynamic recrystallization mechanism of an AZ31 Mg alloy under high-throughput gradient thermal compression. *Mater. Sci. Eng. A* **2022**, *847*, 143338. [CrossRef]
45. Ullmann, M. Rekristallisationsverhalten von Geglühtem AZ31-Gießwalzband beim Warmwalzen. Ph.D. Thesis, TU Bergakademie Freiberg, Freiberg, Germany, 2014.
46. Poletti, C.; Dieringa, H.; Warchomicka, F. Local deformation and processing maps of as-cast AZ31 alloy. *Mater. Sci. Eng. A* **2009**, *516*, 138–147. [CrossRef]
47. Wang, T.; Nie, K.; Deng, K.; Liang, W. Analysis of hot deformation behavior and microstructure evolution of as-cast SiC nanoparticles reinforced magnesium matrix composite. *J. Mater. Res.* **2016**, *31*, 3437–3447. [CrossRef]
48. Arndt, F.; Berndorf, S.; Moses, M.; Ullmann, M.; Prahl, U. Microstructure and Hot Deformation Behaviour of Twin-Roll Cast AZ31 Magnesium Wire. *Crystals* **2022**, *12*, 173. [CrossRef]
49. Sherby, O.D.; Taleff, E.M. Influence of grain size, solute atoms and second-phase particles on creep behavior of polycrystalline solids. *Mater. Sci. Eng. A* **2002**, *322*, 89–99. [CrossRef]
50. Peng, W.P.; Li, P.J.; Zeng, P.; Lei, L.P. Hot deformation behavior and microstructure evolution of twin-roll-cast Mg–2.9Al–0.9Zn alloy: A study with processing map. *Mater. Sci. Eng. A* **2008**, *494*, 173–178. [CrossRef]

51. Liu, Z.; Xing, S.; Bao, P.; Li, N.; Yao, S.; Zhang, M. Characteristics of hot tensile deformation and microstructure evolution of twin-roll cast AZ31B magnesium alloys. *Trans. Nonferrous Met. Soc. China* **2010**, *20*, 776–782. [CrossRef]
52. Xu, S.W.; Matsumoto, N.; Kamado, S.; Honma, T.; Kojima, Y. Effect of Mg17Al12 precipitates on the microstructural changes and mechanical properties of hot compressed AZ91 magnesium alloy. *Mater. Sci. Eng. A* **2009**, *523*, 47–52. [CrossRef]

Disclaimer/Publisher's Note: The statements, opinions and data contained in all publications are solely those of the individual author(s) and contributor(s) and not of MDPI and/or the editor(s). MDPI and/or the editor(s) disclaim responsibility for any injury to people or property resulting from any ideas, methods, instructions or products referred to in the content.

Article

Structure, Phase Composition, and Mechanical Properties of ZK51A Alloy with AlN Nanoparticles after Heat Treatment

Anastasia A. Akhmadieva *, Anton P. Khrustalev, Mikhail V. Grigoriev, Ilya A. Zhukov and Alexander B. Vorozhtsov

Faculty of Physics and Engineering, National Research Tomsk State University, 36 Lenin Ave, 634050 Tomsk, Russia; tofik0014@gmail.ru (A.P.K.); mvgrigoriev@yandex.ru (M.V.G.); gofra930@gmail.com (I.A.Z.); abv1953@mail.ru (A.B.V.)
* Correspondence: nas99.9@yandex.ru

Abstract: The paper studies the influence of the content of aluminum nitride nanoparticles on the structure and mechanical properties of the ZK51A magnesium alloy. The microstructure investigations with optical and electron microscopy show that 1 wt.% AlN promotes the best grain refinement and size distribution. According to tensile strength testing of the ZK51A alloy, grain refinement is not a dominating mechanism in the property improvement of the alloy after heat treatment. The maximum values of mechanical parameters are achieved at the lowest (0.1 wt.%) content of aluminum nitride. The main mechanism of mechanical characteristics increase with the addition of AlN nanoparticles is dispersion hardening.

Keywords: magnesium alloys; ZK51A; nanoparticles; aluminum nitride

1. Introduction

Magnesium alloys are considered to be new-generation materials for the space, aviation, and automobile industries due to their processing and operating parameters, low density, machinability, and specific strength [1–3]. Their significant shortcomings include low formability, insufficient mechanical strength, and corrosion resistance. The creation of cast magnesium alloys with improved strength properties can extend the range of their application [4,5].

The Mg–Zn–Zr system alloy ZK51A is one of the promising magnesium casting alloys. The main hardening mechanism of this alloy is a solid solution associated with the solubility of alloying components (Zn, Zr, Cd) in magnesium [6,7]. The presence of zirconium in this alloy provides its fine-grained structure, as this element is one of the most efficient grain refiners in aluminum-free magnesium alloys [8,9]. This is conditioned by the hexagonal close-packed (HCP) structure of both zirconium and magnesium, and zirconium particles act as magnesium nuclei, which generate the additional solidification centers and limit the grain growth during curing, and thus melt in liquid magnesium [10,11].

The structure of the casting alloy is a solid solution of zinc (Z) and zirconium (Zr) in magnesium (Mg). The MgZn intermetallic phase is located along Mg grain boundaries. After heat treatment, the MgZn phase and Zr-based solid solution sediment form the oversaturated solid solution, thereby hardening it [12,13]. In works [14–17], heat treatment of the ZK51A alloy is performed in the mode T1 (300 °C annealing for 6 h followed by cooling in air).

There are methods of increasing mechanical properties of Mg alloys by dispersion hardening with refractory nano- and microparticles [18–21]. In [22–24], aluminum nitride (AlN) is used as a promising compound for hardening Mg alloys, since it has the HCP crystal lattice with a = 0.312 nm and c = 0.4988 nm parameters almost similar to those of the Mg matrix, i.e., a = 0.3209 nm and c = 0.5211 nm [25,26]. AlN nanoparticles used as reinforcing particles are very interesting due to the high specific strength and low

Citation: Akhmadieva, A.A.; Khrustalev, A.P.; Grigoriev, M.V.; Zhukov, I.A.; Vorozhtsov, A.B. Structure, Phase Composition, and Mechanical Properties of ZK51A Alloy with AlN Nanoparticles after Heat Treatment. *Metals* **2024**, *14*, 71. https://doi.org/10.3390/met14010071

Academic Editor: Ruizhi Wu

Received: 7 October 2023
Revised: 25 December 2023
Accepted: 26 December 2023
Published: 8 January 2024

Copyright: © 2024 by the authors. Licensee MDPI, Basel, Switzerland. This article is an open access article distributed under the terms and conditions of the Creative Commons Attribution (CC BY) license (https://creativecommons.org/licenses/by/4.0/).

coefficient of thermal expansion, high melting point, and hardness [27,28]. Also, due to the small addition of particles, the cost value of alloys grows insignificantly. Research in this field shows that the incorporation of nonmetal microparticles [23,24,29] and nanoparticles [21,22,30] in the Mg matrix improves its strength properties and refines its grain structure. At the same time, very little published information is available concerning the integrated effect of heat treatment and nanoparticle content in the Mg matrix, including alloys of the Mg–Zn–Zr system. Casting is the most common production method of Mg alloys due to the potentiality of creating diverse shapes, high performance, large-lot production, and high-quality surface of the obtained products [31,32].

However, the use of nanoparticles in the production of casting alloys is associated with problems of agglomeration and flotation, which result in the high porosity of alloys. These problems can be avoided via the introduction of nanoparticles consisting of the master alloy during mechanical [33], ultrasonic [34], and vibration [35] treatment of the melt. Despite a large number of studies showing the positive effect of aluminum nitride nanoparticles on the structure and properties of magnesium alloys, the effect of their content in the metal matrix has still not been sufficiently studied.

The aim of this work is to investigate the effect of heat treatment and dispersion hardening on the ZK51A magnesium alloy with different contents of AlN nanoparticles (from 0.1 to 1 wt.%) on its structure, physical and mechanical properties, and fracture.

2. Materials and Methods

2.1. Casting Magnesium Alloys

Magnesium alloy ZK51A consisting of 93.58 to 95.4% Mg, 4 to 5% Zn and 0.6 to 1.1% Zr produced by a conductor electric explosion in nitrogen [36] and machine-milled magnesium powder MPF-4 were used as initial materials. The AlN nanopowder was preliminary deagglomerated and cleaned ultrasonically in ethyl alcohol for 10 min using an Ultrasonic Cleaner DK-300S. The mass ratio of the AlN nanopowder and ethyl alcohol was 1:5. The micropowder MPF-4 was added to the obtained slurry in the amount of 95 wt.% relative to the AlN content and then ultrasonically treated for 10 min with simultaneous stirring in an ULAB US-2200A overhead stirrer rotating at 160 rpm for 10 min. The powder mix Mg–5 wt.% AlN was dried in a vacuum oven at 70 °C. The obtained powder mix was compressed into tablets using a hydraulic press capable of applying a 4 t load with a steel mold with a working area of 40 mm (Figure 1).

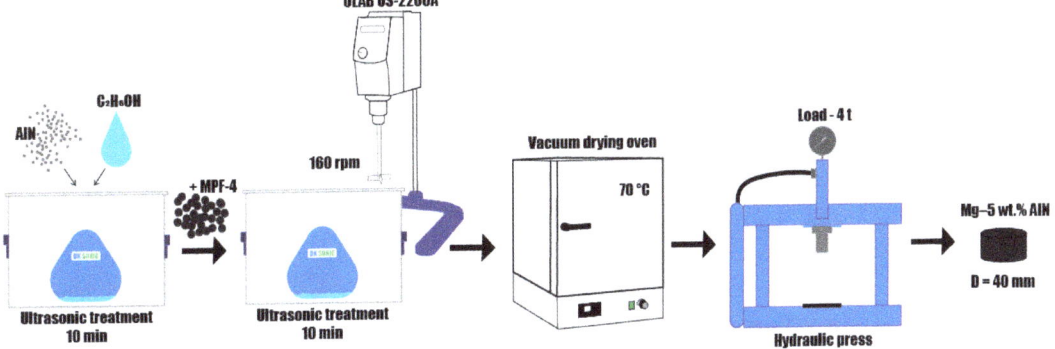

Figure 1. Scheme of obtaining the master alloy.

A total of 2000 g of the ZK51A alloy was placed in the original steel crucible [37] and heated up to 700 °C at a constant argon blow. Then, the melt was mixed for 30 s and introduced in the Mg–AlN master alloy at 690 °C and mixed again at 500 rpm for 1 min. The AlN nanoparticle content in the melt was 0.1, 0.5, and 1 wt.%. At 660 °C, the melt was poured into the steel crucible with a diameter of 35 mm and a height of 200 mm and subjected to vibrations at 0.5 mm amplitude and 60 Hz frequency until complete melt solidification. The initial ZK51A alloy was fabricated with the same parameters without the introduction of the master alloy Mg–5 wt.% AlN. Heat treatment in the mode T1 was performed in a muffle oven at 170 °C, ageing for 22 h followed by furnace cooling (Figure 2).

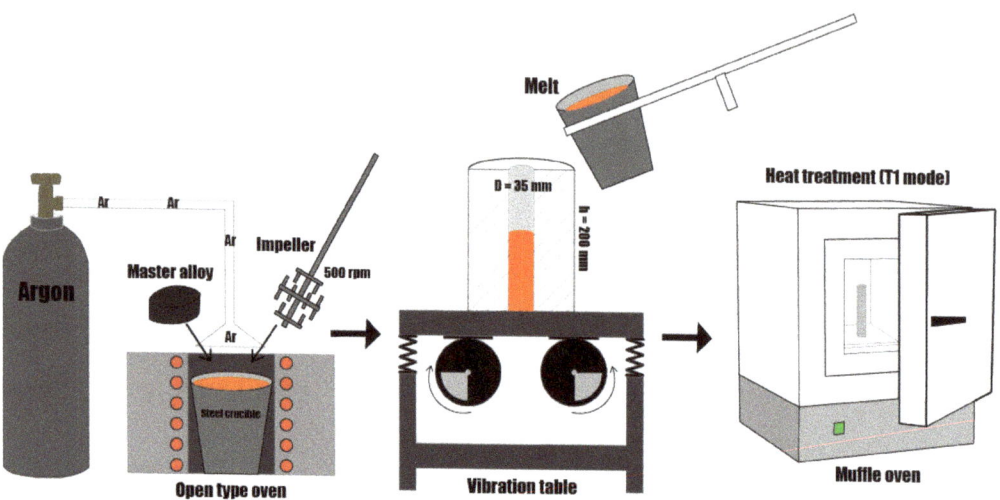

Figure 2. Schematic drawing of the process of the production of alloys.

2.2. Characterization

Initial AlN nanoparticles were investigated on a PHILIPS CM30 Scanning Transmission Electron Microscope (TEM) (Koninklijke Philips N.V., Amsterdam, The Netherlands) using a tungsten cathode. MIRA 3 LMU (Tescan Orsay Holding, Brno, Czech Republic) scanning electron microscope and Olympus GX71 inverted metallurgical microscope (Olympus Scientific Solutions Americas, Waltham, MA, USA) were used to investigate the fine structure of the obtained materials. The alloy surface was etched in picric acid ($C_6H_2(NO_2)_3OH$). X-ray diffraction patterns were recorded on a Shimadzu XRD-6000 Diffractometer (Shimadzu, Kyoto, Japan). Uniaxial tension tests were carried out on $25 \times 6 \times 2$ mm plate-like specimens at a strain rate of 0.001 s^{-1} at room temperature using an Instron 3369 Dual Column Tabletop Testing System (London, UK). Brinell and Vickers hardness testers, Metolab 701 and Metolab 503 (Moscow, Russia), were used to determine the hardness and microhardness, respectively. To measure Brinell hardness, a spherical indenter with a radius of 2.5 mm was used with a force of 62.5 kg and an exposure time of 30 s; the size of the indentations was controlled in the range of 0.2 D < d < 0.6 D. Microhardness testing was carried out on a Metolab 502 microhardness tester using the Vickers method with an indenter in the form of a diamond pyramid (base angle 136°) with a load on the indenter of 50 g and a dwell time of 20 s.

3. Results and Discussion

Figure 3a presents the TEM image of AlN nanoparticles obtained by the conductor electric explosion. The average particle size is 83 nm (Figure 3b). The powder also consists of 200 μm particle agglomerates. The elemental composition of this powder is presented in Table 1.

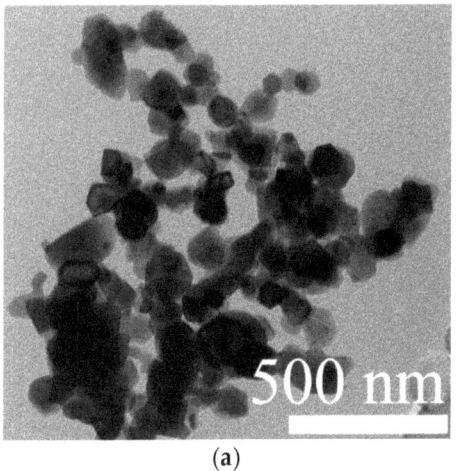 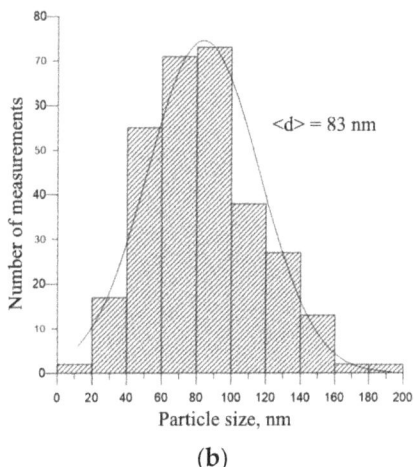

Figure 3. A TEM image of the AlN powder (**a**) and a block diagram of the particle distribution (**b**), where <d> is the average particle size.

Table 1. Elemental composition of AlN powder, wt.%.

Al	N	Si	C	S	Fe	O	Cl	Cu	Ni	P
66.8621	31.6132	0.1524	0.0561	0.0013	0.0354	1.1026	0.1178	0.0217	0.0192	0.0182

According to the X-ray diffraction (XRD) analysis, the initial ZK51A alloy consists of α-Mg with lattice parameters a = 0.32040 nm and c = 0.52005 nm, which correlates with the data from [38,39]. The phase composition analysis of the alloy specimens with nanoparticles shows the presence of aluminum nitride with the content varying in a wide range that can be attributed to the nonuniform particle distribution and a small scan area. The size of the coherent scattering region (CSR) of the AlN phase is within the range of the measured powder particle size. Heat treatment does not affect the SCR size of AlN nanoparticles, although for the solid solution Mg–Zn, it significantly grows. Figure 4 and Table 2 show parameters of phases and the structure of synthesized alloys.

In Figure 5a–d, the optical images demonstrate the ZK51A–AlN alloy microstructure at the center of the sample after etching using a polarizing filter. The average grain size of the initial alloy (Figure 5a) is 46 μm, while at the center and edge of the ingot, it is 69 and 35 μm, respectively, which is associated with the rapid cooling near the wall of the chill mold. In the case of 0.1 wt.% AlN (Figure 5b), the grain boundaries are well-defined, the average grain size is 53 μm, and grains become larger towards the ingot center. In the case of 0.5 (Figure 5c) and 1 wt.% AlN (Figure 5d), the structure is homogeneous, with the average grain size of 46 and 54 μm, respectively. The grain refinement is probably conditioned by the formation of additional solidification centers due to the incorporation of AlN nanoparticles uniformly distributed throughout the melt [23].

Optical images in Figure 6 show the ZK51A alloy microstructure after polishing before etching, before and after heat treatment, and after the addition of AlN nanoparticles. The microstructure without nanoparticles has dark elongated inclusions distributed in the ingot volume, which are probably intermetallic compounds of the Mg–Zn system located at the grain boundaries [17,40]. Spherical inclusions in the grain body are identified as the Zn–Zr phase.

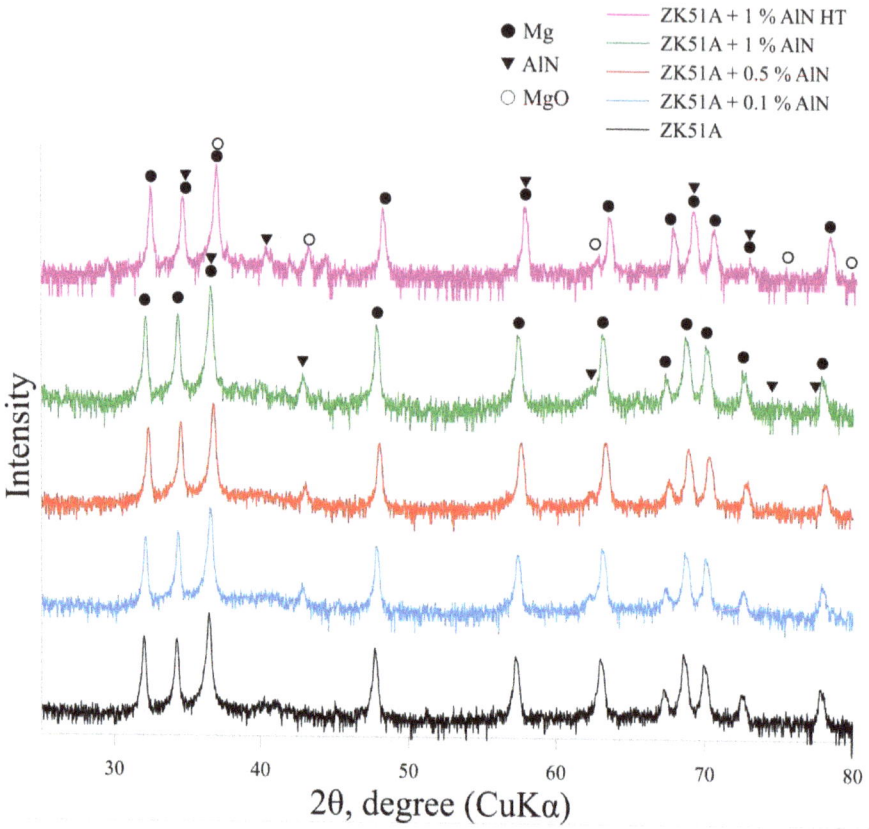

Figure 4. XRD patterns of synthesized alloys.

Table 2. Parameters of phases and the structure of synthesized alloys.

Alloy Specimens	Phases	Phase Content, wt.%	Lattice Parameters, Å	CSR Size, nm	$\Delta d/d \cdot 10^{-3}$
ZK51A	α-Mg	100	a = 3.2040 c = 5.2005	57	0.8
ZK51A + 0.1 wt.% AlN	α-Mg	88	a = 3.2038 c = 5.2006	65	0.2
	AlN	12	a = 4.2324	39	1.1
ZK51A + 0.5 wt.% AlN	α-Mg	97	a = 3.2028 c = 5.1977	83	1.4
	AlN	3	a = 4.2018	10	4.8
ZK51A + 1 wt.% AlN	α-Mg	95	a = 3.2058 c = 5.2046	84	0.8
	AlN	5	a = 4.2634	27	0.8
ZK51A + 1 wt.% AlN (HT)	α-Mg	97	a = 3.2048 c = 5.2091	319	1.3
	AlN	2	a = 4.5003	45	1.9

The MgZn intermetallic content of the initial alloy after heat treatment considerably grows and is located nearby the ingot edges. After the addition of AlN nanoparticles, a lot of defects appear in the structure, and the MgZn and ZnZr phases remain. The MgZn intermetallic creates 'filaments' linking to each other. The alloy structure consisting of 1% of nanoparticles, is characterized by darker grain-style inclusions throughout the ingot before and after heat treatment. Perhaps, either etching during polishing, or a large number of nanoparticles result in the formation of the new phase. After heat treatment, the microstructure does not significantly change with the nanoparticle addition.

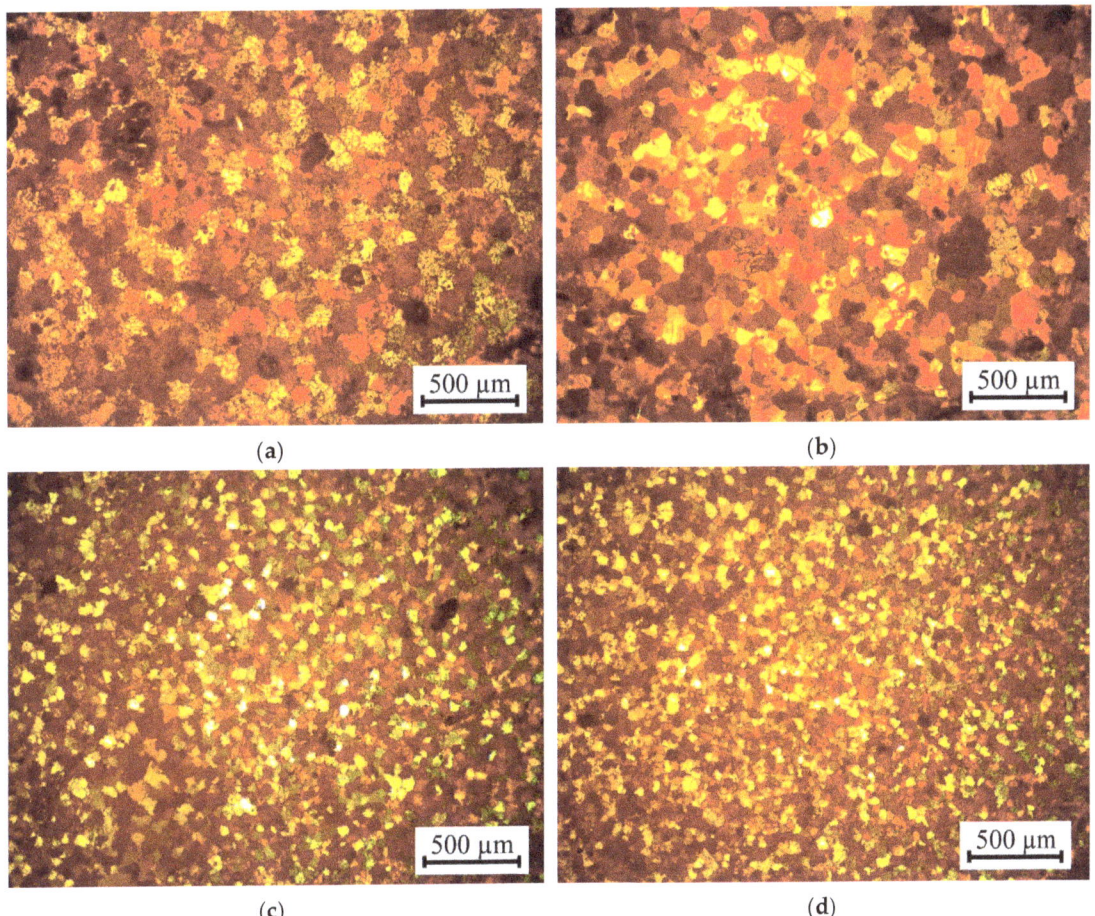

Figure 5. Optical images of the alloy microstructure with different AlN content: (**a**) ZK51A, (**b**) ZK51A + 0.1 wt.% AlN, (**c**) ZK51A + 0.5 wt.% AlN, and (**d**) ZK51A + 1 wt.% AlN.

SEM images in Figure 7 demonstrate the alloy microstructure with 1% of AlN nanoparticles before and after heat treatment. One can see that the microstructure does not differ in either states of the alloy. It consists of equiaxial grains with the average size of 54 μm. The MgZn intermetallic in the Mg–Zn–Zr system alloy appears along the grain boundaries and serves as a grain-growth constraint [13,16]. The ZnZr phase inclusions are not found as, in theory, they appear in the grain body. The cross-sectional analysis does not show inclusions on the surface. The dark areas in the SEM images represent pores formed due to imperfect casting technology.

SEM in combination with energy dispersive X-ray spectroscopy (SEM-EDS) in Figure 8, shows large dark areas identified as fragments of the Mg–AlN master alloy. In addition to the MgZn phase, the intergranular space consists of the Zr-containing phase. According to [9,16], the Mg–Zn–Zr system cast alloys consist of 4.5 to 8.5% of zinc and 0.8 to 1.0% of zirconium, and the latter generates Zn_2Zr_3, $ZnZr_2$, and Zn_2Zr phases. Depending on the heat treatment conditions, the region with ZnZr intermetallic formations can change. Figure 9 presents the elemental composition of inclusions in the ZK51A + 1% AlN alloy after heat treatment. Heat treatment promotes homogenization of the alloy microstructure due to diffusion interaction between the elements of the undissolved master alloy and the magnesium matrix.

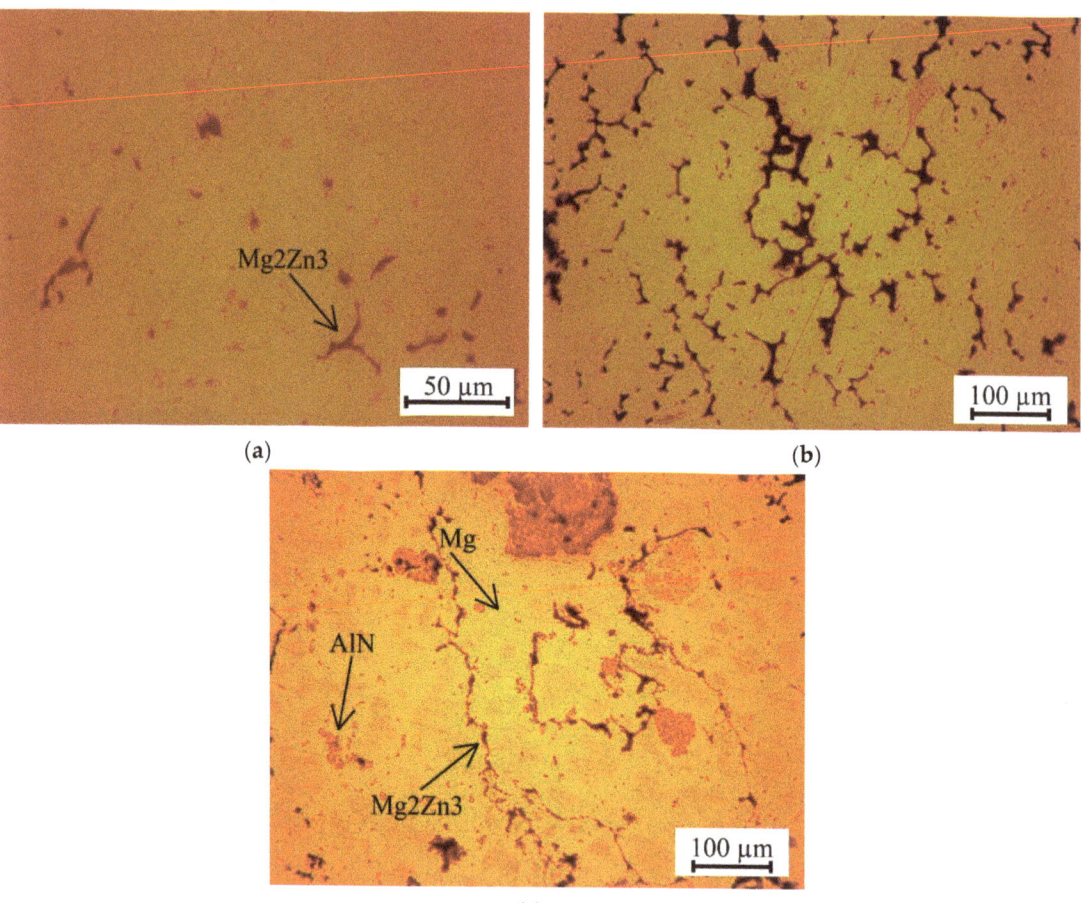

Figure 6. ZK51A alloy microstructure: (**a**)—without AlN nanoparticles, (**b**)—after heat treatment in mode T1, and (**c**)—with 1 wt.% AlN before heat treatment.

Table 3 summarizes measurement results of the hardness and density of the alloys obtained. Figure 10 contains plots of the grain size and hardness relative to the AlN content in cast alloys and alloys after heat treatment. One can see that the hardness lowers from 55 to 47 HB with the nanoparticle content increasing from 0.1 to 1 wt.%, regardless of heat treatment.

Dependence of the grain size and porosity of the AlN content in cast and heat-treated alloys are presented in Figure 11. The average porosity does not exceed 2.5%, while its lowest value (0.5%) is gained after the addition of 0.1 wt.% AlN. When the AlN content increases up to 0.5 and 1 wt.%, the porosity reaches 1.3 and 2.5%, respectively. This can be attributed to additional gases penetrating in the melt during the nanoparticle introduction.

The microhardness AlN content dependence on the AlN content are given in Figure 12 for the alloy before and after heat treatment. It is found that heat treatment has no effect on the microhardness, while the increased nanoparticle content results in its reduction, i.e., 62 and ~50 HV for the initial alloy and the alloy with 1 wt.% AlN, respectively.

Table 3. Parameters of ZK51A–AlN cast alloys and alloys after heat treatment.

Alloys	Hardness, HB	Density, g/cm³	Grain Size, μm
ZK51A	56 ± 2	1.79 ± 0.1	46 ± 26
ZK51A (HT)	59 ± 1	1.79 ± 0.1	
ZK51A + 0.1 wt.% AlN	54 ± 3	1.8 ± 0.1	53 ± 29
ZK51A + 0.1 wt.% AlN (HT)	55 ± 1	1.8 ± 0.1	
ZK51A + 0.5 wt.% AlN	51 ± 2	1.8 ± 0.1	46 ± 18
ZK51A + 0.5 wt.% AlN (HT)	52 ± 2	1.79 ± 0.1	
ZK51A + 1 wt.% AlN	44 ± 4	1.78 ± 0.1	54 ± 17
ZK51A + 1 wt.% AlN (HT)	50 ± 3	1.77 ± 0.1	

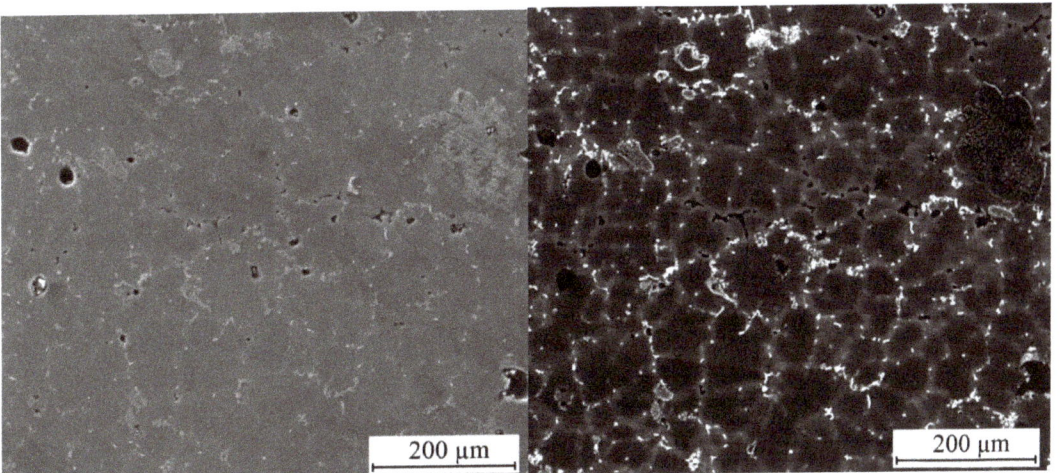

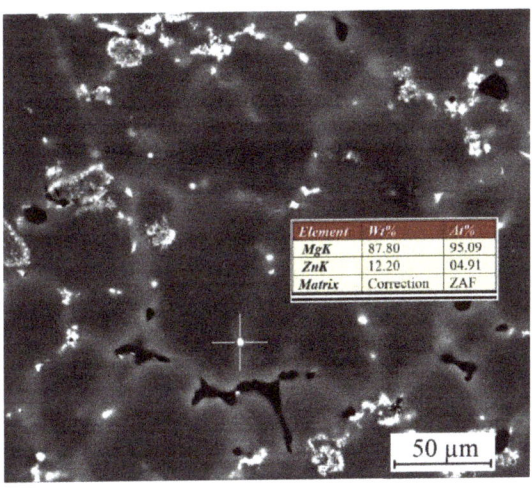

(a)

Figure 7. *Cont.*

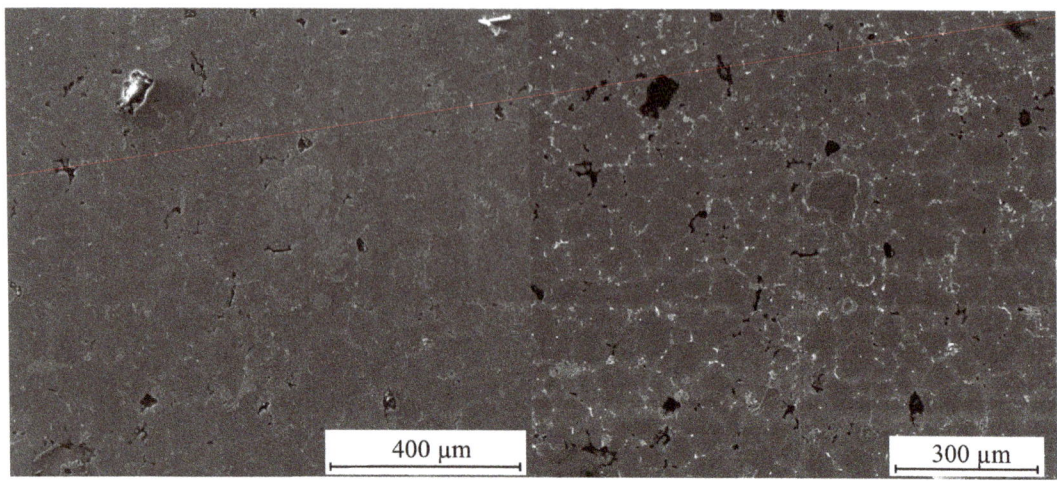

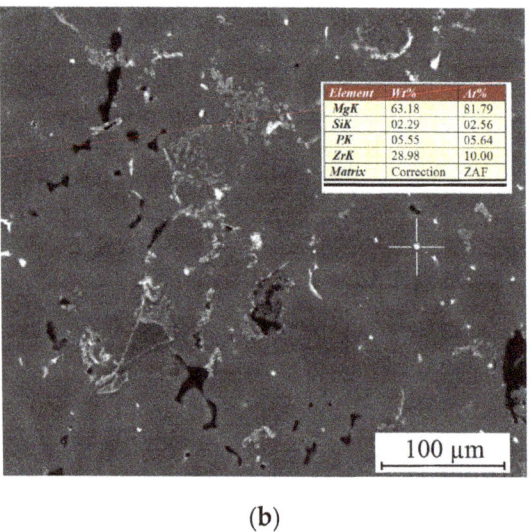

(b)

Figure 7. SEM images of ZK51A + 1% AlN alloy structure before (**a**) and after (**b**) heat treatment. White crosses indicate the area of elemental analysis.

Figure 13 presents the stress–strain curves for the obtained alloys after heat treatment. As can be seen from Table 4, the yield strength, tensile strength, and plasticity of the initial alloy without nanoparticles are 63 MPa, 151 MPa, and 6.4%, respectively. After the addition of 0.1, 0.5, and 1 wt.% AlN nanoparticles, the yield strength grows up to 72, 67, and 86 MPa, respectively. The highest growth in the yield strength and plasticity relative to the initial allo, is observed at 0.1 wt.% AlN, viz. from 151 to 212 MPa and from 6.4 to 19.7%, respectively.

The main mechanism of the mechanical characteristics increase with the addition of AlN nanoparticles is dispersion hardening. It is suggested that the introduction of particles into the alloy structure can lead to the deflection of a potential crack from the grain boundary into its volume, as well as greater involvement of the metal matrix in the deformation and fracture process [41,42]. These results are in agreement with earlier studies on the introduction of AlN particles into the AZ91 alloy [37]. However, this effect is achieved by adding particles in an amount not exceeding 0.1 wt.%. As the number of particles increases, agglomerates are formed, which do not maximize mechanical properties and result in additional fracture centers.

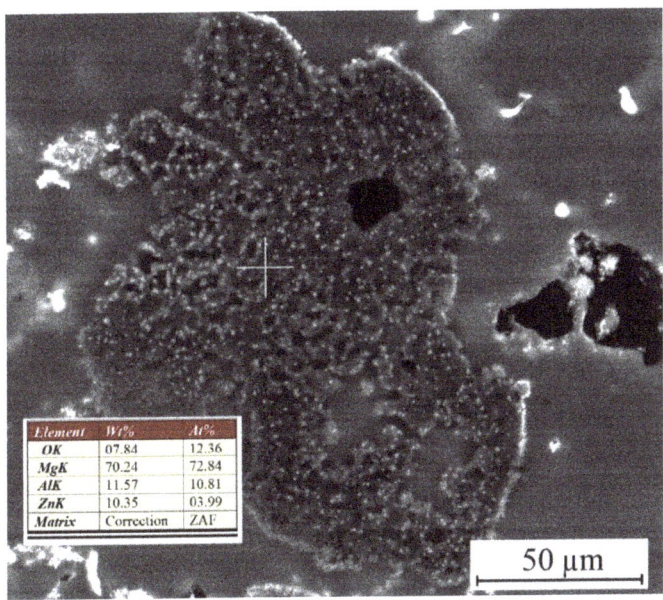

Figure 8. SEM-EDS of master alloy fragments in ZK51A + 1% AlN alloy before heat treatment. A white cross indicates an elemental analysis area.

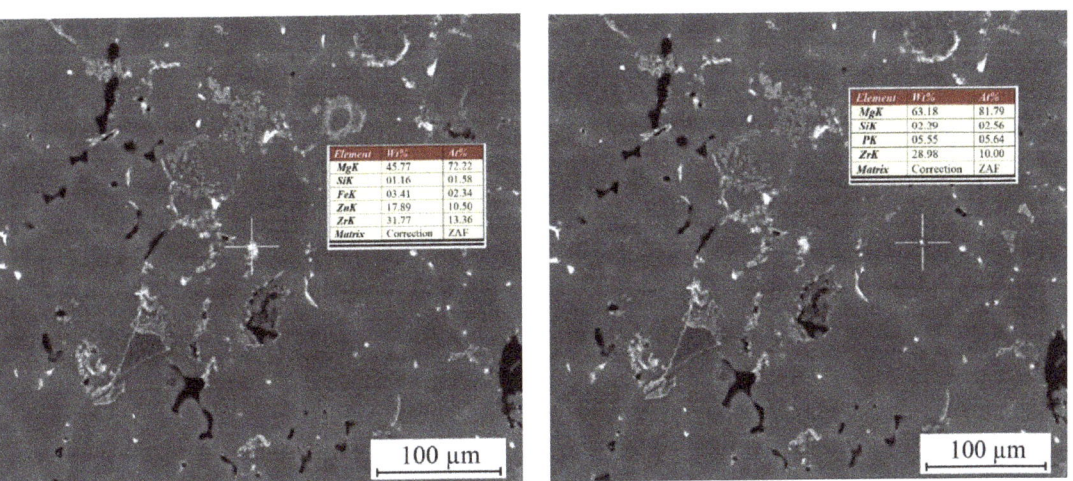

Figure 9. Elemental composition of inclusions in the ZK51A + 1% AlN alloy after heat treatment. White crosses indicate the area of elemental analysis.

The increase in yield strength values of the alloy occurs in accordance with the Hall-Petch law, according to which $\sigma_{0.2}$ increases when a uniform fine-grained structure is achieved [43].

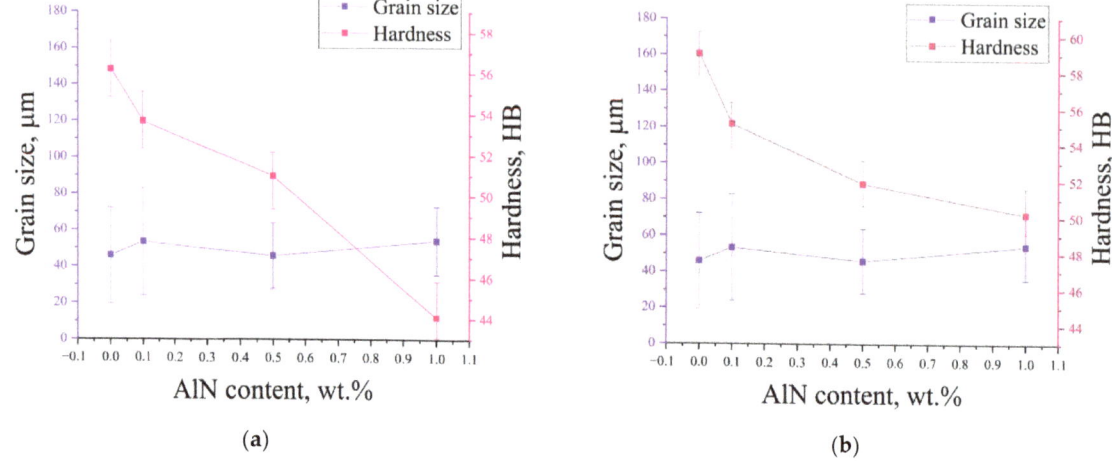

Figure 10. Dependence of grain size and hardness of AlN content in cast (**a**) and heat-treated (**b**) alloys.

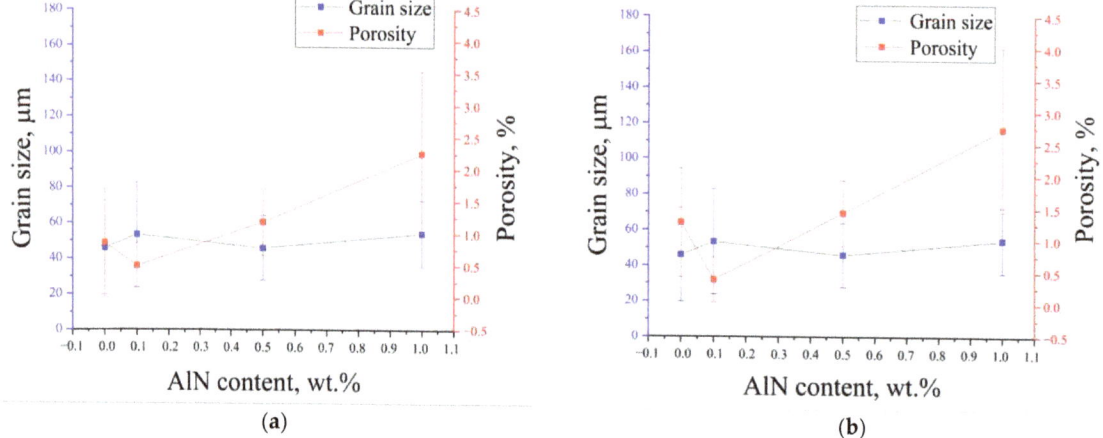

Figure 11. Dependence of grain size and porosity on AlN content in cast (**a**) and heat-treated (**b**) alloys.

SEM observations of the fracture surface after tensile strength testing show a ductile transgranular fracture mechanism for all alloys. SEM images of the fracture surface of heat-treated alloys are presented in Figure 14a–d. There is a particle coalescence comprising aluminum (see Figure 15). These results would suggest that although introduced non-uniformly, AlN nanoparticles are present in the structure and thus some of them cannot be identified due to their nanoscale size.

Using magnification in Figure 16, we identified ~5 μm inclusions on the fracture surface of heat-treated ZK51A + 1 wt.% AlN alloy, which consisted of zirconium. Supposedly, these inclusions were the ZnZr phase, which, in theory, released inside the grain in the form of coarse particles [16].

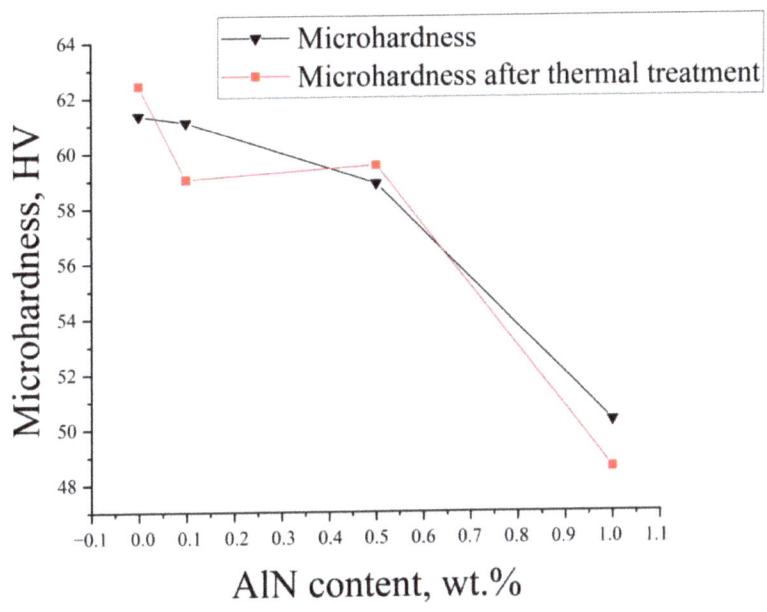

Figure 12. Microhardness dependence on AlN content before and after heat treatment.

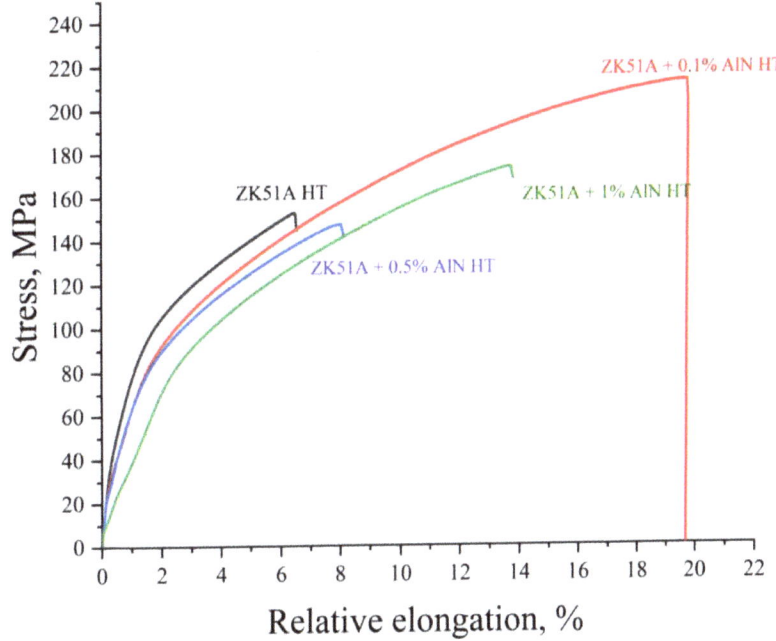

Figure 13. Stress–strain curves of alloys after heat treatment.

Table 4. Mechanical properties of ZK51A–AlN cast alloys and alloys after heat treatment.

Alloys	$\sigma_{0.2}$, MPa	σ, MPa	ε, %	Microhardness, HV
ZK51A (HT)	63 ± 4	151 ± 8	6.4 ± 0.4	62.5 ± 9.2
ZK51A + 0.1 wt.% AlN (HT)	72 ± 7	212 ± 11	19.7 ± 0.6	59 ± 9.6
ZK51A + 0.5 wt.% AlN (HT)	67 ± 5	148 ± 5	8 ± 0.7	59.5 ± 9.4
ZK51A + 1 wt.% AlN (HT)	86 ± 5	174 ± 6	13.8 ± 0.3	48.6 ± 6.1

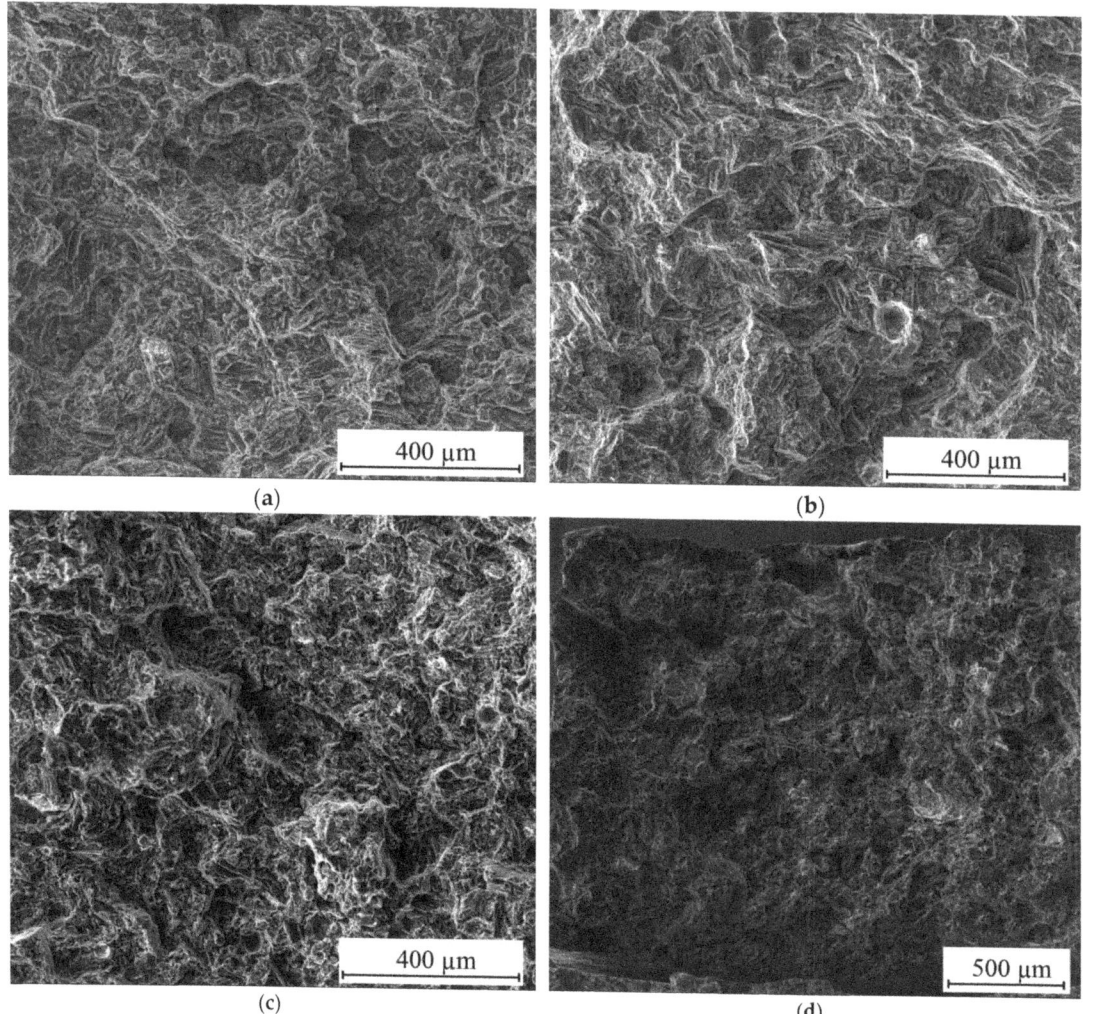

Figure 14. SEM images of the ductile fracture surface of heat-treated alloys failed after tensile strength testing: (a) ZK51A, (b) ZK51A + 0.1 wt.% AlN, (c) ZK51A + 0.5 wt.% AlN, and (d) ZK51A + 1 wt.% AlN.

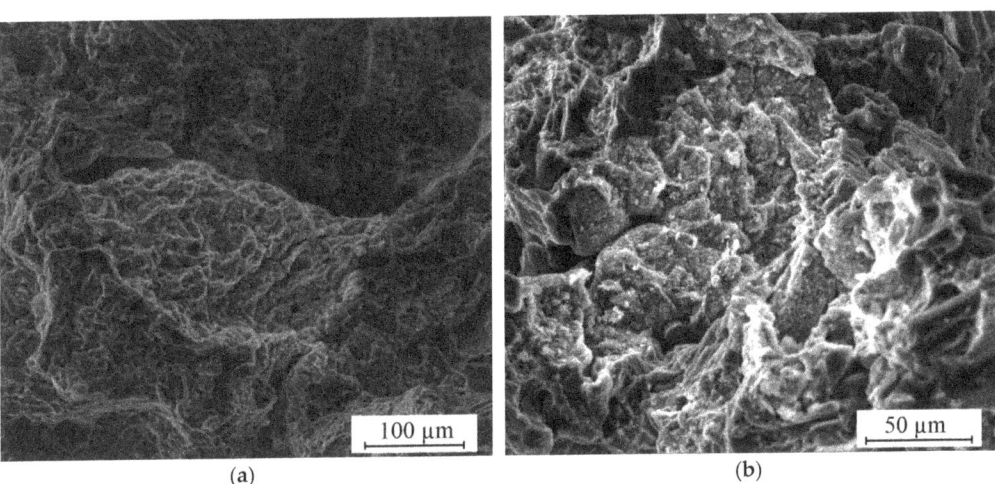

Figure 15. Fracture surface of heat-treated ZK51A + 0.5 wt.% AlN (**a**) and ZK51A + 1 wt.% AlN (**b**) alloys.

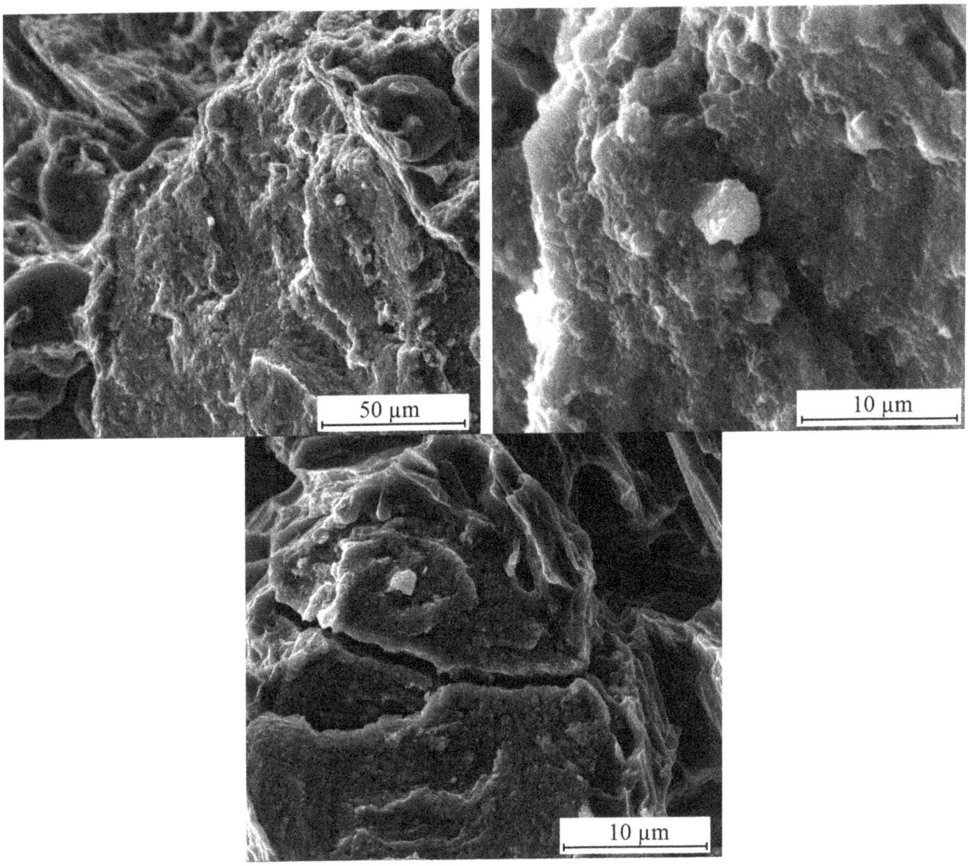

Figure 16. ZnZr phase (light inclusions) in the grain body of ZK51A + 1% AlN alloy after heat treatment.

4. Conclusions

Summing up the results, it can be concluded that the AlN nanoparticle content of 0.5 and 1 wt.% provided the homogeneous grain distribution in Mg alloys having the grain size of 46 and 54 μm, respectively. That was due to the formation of additional solidification centers. At the same time, heat treatment did not affect the grained structure of the alloy.

Heat treatment promotes homogenization of the microstructure of magnesium alloys containing aluminum nitride particles, which reduces the negative influence of undissolved components of the Mg–AlN master alloy.

It was found that the incorporation of 0.1 wt.% AlN resulted in a simultaneous increase in the yield strength (63 to 72 MPa), tensile strength (151 to 212 MPa), and plasticity (6.4 to 19.7%) of the heat-treated alloy due to its dispersion hardening. A further growth in the nanoparticle content led to a reduction in mechanical properties of the alloy because of the formation of agglomerates hampering the attainment of the best mechanical properties. Despite the grain refinement and increase in the mechanical characteristics of the magnesium alloy during tension, its hardness decreases with increasing aluminum nitride content, since the introduction of particles contributes to the formation of additional pore space.

Aluminum nitride particles do not affect the mechanism of destruction of the metal matrix, which is characterized by a lamellar transcrystalline nature of destruction.

Author Contributions: Conceptualization, A.B.V., I.A.Z. and A.P.K.; methodology, A.B.V., I.A.Z., M.V.G. and A.P.K.; investigation, A.A.A., M.V.G. and A.P.K.; writing—original draft preparation, A.A.A. and A.P.K.; writing—review and editing, A.A.A., A.P.K. and I.A.Z.; project administration and funding acquisition, I.A.Z.; and supervision, A.B.V. All authors have read and agreed to the published version of the manuscript.

Funding: This work was carried out with financial support from the Ministry of Education and Science of the Russian Federation (State assignment No. FSWM-2020-0028).

Data Availability Statement: The data presented in this study are available in the article.

Acknowledgments: The research was done using equipment of the Tomsk Regional Core Shared Research Facilities Centre of National Research Tomsk State University.

Conflicts of Interest: The authors declare no conflicts of interest.

References

1. Song, G.L.; Xu, Z.Q. The surface, microstructure and corrosion of magnesium alloy AZ31 sheet. *Electrochim. Acta* **2010**, *55*, 4148–4161. [CrossRef]
2. Morozova, G.I. Phase composition and corrosion resistance of magnesium alloys. *Met. Sci. Heat. Treat.* **2008**, *50*, 100–104. [CrossRef]
3. Jayasathyakawin, S.; Ravichandran, M.; Baskar, N.; Anand Chairman, C.; Balasundaram, R. Mechanical properties and applications of Magnesium alloy–Review. *Mat. Tod. Proc.* **2020**, *27*, 909–913. [CrossRef]
4. Penghuai, F.; Liming, P.; Haiyan, J.; Wenjiang, D.; Chunquan, Z. Tensile properties of high strength cast Mg alloys at room temperature: A review. *China Foundry* **2014**, *11*, 277–286.
5. Ramalingam, V.V.; Ramasamy, P.; Kovukkal, M.D.; Myilsamy, G. Research and Development in Magnesium Alloys for Industrial and Biomedical Applications: A Review. *Met. Mater. Int.* **2020**, *26*, 409–430. [CrossRef]
6. Reinor, G.V. *The Physical Metallurgy of Magnesium and Its Alloys*; Metallurgiya: Moscow, Russia, 1964; 477p. (In Russian)
7. Li, P.; Hou, D.; Han, E.H.; Chen, R.; Shan, Z. Solidification of Mg–Zn–Zr alloys: Grain growth restriction, dendrite coherency and grain size. *Acta Metall. Sin. Engl. Lett.* **2020**, *33*, 1477–1486. [CrossRef]
8. Qian, M.; Das, A. Grain refinement of magnesium alloys by zirconium: Formation of equiaxed grains. *Scr. Mater.* **2006**, *54*, 881–886. [CrossRef]
9. Xing, F.; Guo, F.; Su, J.; Zhao, X.; Cai, H. The existing forms of Zr in Mg-Zn-Zr magnesium alloys and its grain refinement mechanism. *Mater. Res. Express.* **2021**, *8*, 066516. [CrossRef]
10. Song, C.; Han, Q.; Zhai, Q. Review of grain refinement methods for as-cast microstructure of magnesium alloy. *China Foundry* **2009**, *6*, 93–103.
11. Yang, W.; Liu, L.; Zhang, J.; Ji, S.; Fan, Z. Heterogeneous nucleation in Mg–Zr alloy under die casting condition. *Mater. Lett.* **2015**, *160*, 263–267. [CrossRef]
12. Koltygin, A.V.; Bazhenov, V.E.; Letyagin, N.V.; Belov, V.D. The influence of composition and heat treatment on the phase composition and mechanical properties of ML19 magnesium alloy. *Russ. J. Non-Ferr. Met.* **2018**, *59*, 32–41. [CrossRef]

13. Wang, S.; Zhao, Y.; Guo, H.; Lan, F.; Hou, H. Mechanical and Thermal Conductivity Properties of Enhanced Phases in Mg-Zn-Zr System from First Principles. *Materials* **2018**, *11*, 2010. [CrossRef] [PubMed]
14. Koltygin, A.V.; Bazhenov, V.E. Influence of the chemical composition and heat treatment modes on the phase composition and mechanical properties of the ZK51A (ML12) alloy. *Russ. J. Non-Ferr. Met.* **2018**, *59*, 190–199. [CrossRef]
15. Morozova, G.I.; Tikhonova, V.V.; Lashko, N.F. Phase composition and mechanical properties of cast Mg−Zn−Zr alloys. *Met. Sci. Heat. Treat.* **1978**, *20*, 657–660. [CrossRef]
16. Morozova, G.I.; Mukhina, I.Y. Nanostructural hardening of cast magnesium alloys of the Mg–Zn–Zr system. *Met. Sci. Heat. Treat.* **2011**, *53*, 3. [CrossRef]
17. Mukhina, I.Y. Structure and properties of new foundry magnesium alloys. *Foundry Prod.* **2011**, *12*, 12–14. (In Russian)
18. Vorozhtsov, S.A.; Khrustalyov, A.P.; Eskin, D.G.; Kulkov, S.N.; Alba-Baena, N. The physical-mechanical and electrical properties of cast aluminum-based alloys reinforced with diamond nanoparticles. *Russ. Phys. J.* **2015**, *57*, 1485–1490. [CrossRef]
19. Zheng, H.R.; Li, Z.; You, C.; Liu, D.B.; Chen, M.F. Effects of MgO modified β-TCP nanoparticles on the microstructure and properties of β-TCP/Mg-Zn-Zr composites. *Bioactive Mater.* **2017**, *2*, 1–9. [CrossRef]
20. Wong, W.L.; Gupta, M. Effect of hybrid length scales (micro+ nano) of SiC reinforcement on the properties of magnesium. *Solid State Phenom.* **2006**, *111*, 91–94. [CrossRef]
21. Hassan, S.F.; Gupta, M. Creation of high-performance Mg based composite containing nano-size Al_2O_3 particulates as reinforcement. *J. Metast. Nanocr. Mat.* **2005**, *23*, 151–154. [CrossRef]
22. Zhang, B.; Yang, C.; Zhao, D.; Sun, Y.; Wang, X.; Liu, F. Microstructure characteristics and enhanced tensile properties of in-situ AlN/AZ91 composites prepared by liquid nitriding method. *Mater. Sci. Eng. A* **2018**, *725*, 207–214. [CrossRef]
23. Fu, H.M.; Zhang, M.X.; Qiu, D.; Kelly, P.M.; Taylor, J.A. Grain refinement by AlN particles in Mg–Al based alloys. *J. Alloys Compd.* **2009**, *478*, 809–812. [CrossRef]
24. Cao, G.; Choi, H.; Oportus, J.; Konishi, H.; Li, X. Study on tensile properties and microstructure of cast AZ91D/AlN nanocomposites. *Mater. Sci. Eng. A* **2008**, *494*, 127–131. [CrossRef]
25. Berkmortel, R.; Wang, G.G.; Bakke, P. Fluxless in-house recycling of high purity magnesium die cast alloys at Meridian operations. In Proceedings of the 57th IMA Conference, Vancouver BC, Canada, 21–23 May 2000; pp. 22–27.
26. Liu, P.; Geng, H.R.; Wang, Z.Q.; Zhu, J.R.; Pan, F.S.; Dong, X.B. Effect of AlN on Microstructure and Mechanical Properties of Mg-Al-Zn Alloy. *Mater. Sci. Forum* **2011**, *704–705*, 1095–1099. [CrossRef]
27. Wahab, M.N.; Daud, A.R.; Ghazali, M.J. Preparation and characterization of stir cast-aluminum nitride reinforced aluminum metal matrix composites. *Int. J. Mech. Mater. Eng.* **2009**, *4*, 115–117.
28. Vinayagam, M.; Ravichandran, M. Influence of AlN particles on microstructure, mechanical and tribological behaviour in AA6351 aluminum alloy. *Mater. Res. Express* **2019**, *6*, 106557. [CrossRef]
29. Huang, S.J.; Abbas, A. Effects of tungsten disulfide on microstructure and mechanical properties of AZ91 magnesium alloy manufactured by stir casting. *J. Alloys Compd.* **2020**, *817*, 153321. [CrossRef]
30. Hassan, S.F.; Gupta, M. Enhancing physical and mechanical properties of Mg using nanosized Al_2O_3 particulates as reinforcement. *Metall. Mater. Trans. A* **2005**, *36*, 2253–2258. [CrossRef]
31. Sajuri, Z.B.; Miyashita, Y.; Hosokai, Y.; Mutoh, Y. Effects of Mn content and texture on fatigue properties of as-cast and extruded AZ61 magnesium alloys. *Inter. J. Mech. Sci.* **2006**, *48*, 198–209. [CrossRef]
32. Pan, F.; Yang, M.; Chen, X. A review on casting magnesium alloys: Modification of commercial alloys and development of new alloys. *J. Mater. Sci. Tech.* **2016**, *32*, 1211–1221. [CrossRef]
33. Kakhidze, N.I.; Khrustalev, A.P.; Akhmadieva, A.A.; Zhukov, I.A.; Vorozhtsov, A.B. Influence of tungsten nanoparticles on the structure and mechanical behavior of AA5056 under quasi-static loading. In *Light Metals*; Springer: Cham, Switzerland, 2022; pp. 97–103. [CrossRef]
34. Khrustalyov, A.; Kakhidze, N.; Platov, V.; Zhukov, I.; Vorozhtsov, A. Influence of tungsten nanoparticles on microstructure and mechanical properties of an Al–5% mg alloy produced by casting. *Metals* **2022**, *12*, 989. [CrossRef]
35. Khrustalyov, A.P.; Akhmadieva, A.; Monogenov, A.N.; Zhukov, I.A.; Marchenko, E.S.; Vorozhtsov, A.B. Study of the effect of diamond nanoparticles on the structure and mechanical properties of the medical Mg–Ca–Zn magnesium alloy. *Metals* **2022**, *12*, 206. [CrossRef]
36. Lerner, M.; Vorozhtsov, A.; Guseinov, S.; Storozhenko, P. *Metal Nanopowders: Production Characterization, and Energetic Applications*; Wiley-VCH: Weinheim, Germany, 2014; pp. 79–106.
37. Khrustalyov, A.; Zhukov, I.; Nikitin, P.; Kolarik, V.; Klein, F.; Akhmadieva, A.; Vorozhtsov, A. Study of Influence of aluminum nitride nanoparticles on the structure, phase composition and mechanical properties of AZ91 alloy. *Metals* **2022**, *12*, 277. [CrossRef]
38. Orlov, D.; Pelliccia, D.; Fang, X.; Bourgeois, L.; Kirby, N.; Nikulin, A.Y.; Ameyama, K.; Estrin, Y. Particle evolution in Mg–Zn–Zr alloy processed by integrated extrusion and equal channel angular pressing: Evaluation by electron microscopy and synchrotron small-angle X-ray scattering. *Acta Mater.* **2014**, *72*, 110–124. [CrossRef]
39. Liu, S.; Yang, W.; Shi, X.; Li, B.; Duan, S.; Guo, H.; Guo, J. Influence of laser process parameters on the densification, microstructure, and mechanical properties of a selective laser melted AZ61 magnesium alloy. *J. Alloys Compd.* **2019**, *808*, 151160. [CrossRef]
40. Wang, B.J.; Xu, D.K.; Sun, J.; Han, E.H. Effect of grain structure on the stress corrosion cracking (SCC) behavior of an as-extruded Mg-Zn-Zr alloy. *Corr. Sci.* **2019**, *157*, 347–356. [CrossRef]

41. Dieringa, H.; Katsarou, L.; Buzolin, R.; Szakács, G.; Horstmann, M.; Wolff, M.; Mendis, C.; Vorozhtsov, S.; StJohn, D. Ultrasound Assisted Casting of an AM60 Based Metal Matrix Nanocomposite, Its Properties, and Recyclability. *Metals* **2017**, *7*, 388. [CrossRef]
42. Khrustalyov, A.P.; Garkushin, G.V.; Zhukov, I.A.; Razorenov, S.V.; Vorozhtsov, A.B. Quasi-Static and Plate Impact Loading of Cast Magnesium Alloy ML5 Reinforced with Aluminum Nitride Nanoparticles. *Metals* **2019**, *9*, 715. [CrossRef]
43. Petch, N. The cleavage strength of polycrystals. *J. Iron Steel Inst.* **1953**, *174*, 25–28.

Disclaimer/Publisher's Note: The statements, opinions and data contained in all publications are solely those of the individual author(s) and contributor(s) and not of MDPI and/or the editor(s). MDPI and/or the editor(s) disclaim responsibility for any injury to people or property resulting from any ideas, methods, instructions or products referred to in the content.

Article

Cold Formability of Twin-Roll Cast, Rolled and Annealed Mg Strips

Madlen Ullmann *, Kristina Kittner and Ulrich Prahl

Institute of Metal Forming, Technische Universität Bergakademie Freiberg, Bernhard-von-Cotta-Str. 4, 09599 Freiberg, Germany; kristina.kittner@imf.tu-freiberg.de (K.K.); ulrich.prahl@imf.tu-freiberg.de (U.P.)
* Correspondence: madlen.ullmann@imf.tu-freiberg.de

Abstract: This study investigates the cold formability of twin-roll cast and rolled magnesium strips, specifically focusing on AZ31 and ZAX210 alloys. The aim is to assess the suitability of these alloys for various forming processes. The mechanical properties and formability characteristics of the strips were thoroughly examined to provide insights into their potential applications in transportation industries such as automotive and aerospace. The AZ31 and ZAX210 alloys were subjected to twin-roll casting and rolling processes to produce thin strips. The resulting strips were then evaluated for their cold formability. The results indicate that both alloys exhibit favourable cold formability. The ZAX210 alloy, in particular, demonstrates medium strengths with an average tensile strength of approximately 240 MPa at room temperature. The 0.2% proof stress values range between 136 MPa and 159 MPa, depending on the sampling direction. The total elongation values vary from 28% in the transverse direction to 32% at a 45° angle, indicating excellent ductility. Comparing the two alloys, the AZ31 alloy shows higher strengths due to its higher aluminium content. However, it also exhibits a more pronounced directional dependence of mechanical properties due to the formation of a strong basal texture during hot rolling. The transverse direction experiences reduced total elongation compared to the rolling direction, achieving only about 50% of the total elongation. The average Erichsen Index recorded for AZ31 and ZAX210 strips were 4.9 mm and 7.1 mm, respectively. The ZAX210 strip displays superior formability, which can be attributed to the fine-grained microstructure and the texture softening resulting from the weakening of the basal texture intensity and the splitting of the basal pole towards the rolling direction. In conclusion, the investigated twin-roll cast, rolled and annealed AZ31 and ZAX210 magnesium strips exhibit promising cold formability characteristics. The findings of this study contribute to the understanding of their mechanical behaviour and can guide the selection and optimisation of these alloys for various forming applications.

Keywords: magnesium alloys; cold formability; twin-roll casting; rolling; Mg-Zn-Al-Ca; Erichsen Index; mechanical properties; FLC

Citation: Ullmann, M.; Kittner, K.; Prahl, U. Cold Formability of Twin-Roll Cast, Rolled and Annealed Mg Strips. *Metals* **2024**, *14*, 121. https://doi.org/10.3390/met14010121

Academic Editor: Ruizhi Wu

Received: 1 December 2023
Revised: 10 January 2024
Accepted: 17 January 2024
Published: 19 January 2024

Copyright: © 2024 by the authors. Licensee MDPI, Basel, Switzerland. This article is an open access article distributed under the terms and conditions of the Creative Commons Attribution (CC BY) license (https://creativecommons.org/licenses/by/4.0/).

1. Introduction

Magnesium alloys have gained significant attention in various industries due to their low density, excellent mechanical properties, and high specific strength. However, the inherent poor formability of magnesium alloys at room temperature (RT) poses challenges in their practical applications, particularly in cold forming processes. Sheet metal forming processes like deep drawing, cupping operations or bending operations require elevated temperatures when magnesium alloys are used. To decrease forming temperatures, different factors must be considered. The cold formability of magnesium alloys is influenced by several factors, including alloy composition, microstructure, texture, processing methods and the manufacturing process itself. Among these factors, alloy composition plays a crucial role in determining the formability characteristics [1,2].

The low formability of magnesium alloys at room temperature is mainly attributed to the hexagonal lattice and the strongly pronounced basal texture during conventional wrought processing. This is accompanied by a stronger anisotropy and lower stretch formability. In order to overcome these disadvantages, several studies have focused on increasing room temperature formability by adding alloying elements and developing appropriate processing technologies in order to attain texture weakening resulting in an advantageous effect on the plasticity of wrought magnesium alloys at room temperature [1]. Alloying elements that contribute to the development of weak textures are rare earths elements (REs) and calcium [3–6]. Rare earths elements are disadvantageous due to their rarity, high costs and difficulties in mining, despite their positive property effects. New developed magnesium alloy systems with low RE-content or Ca addition offering a high RT stretch formability include (among others) Mg-Zn-RE, Mg-Sn-RE, Mg-Zn-Ca and Mg-Al-Ca-Mn systems [1,7]. These alloys exhibit an excellent RT formability with a high Index Erichsen (IE) value of between 7 mm and 9 mm. However, increased RT formability is accompanied by lower strength values due to the strength–ductility trade-off dilemma [8]. Magnesium alloys exhibiting higher strength, for example Mg-Mn- or Mg-Al-based alloys, offer inferior formability, with IE values between 3 mm and 5 mm. In turn, the alloying elements RE and Ca have a favourable effect, as they enable an increase in strength through age- or bake-hardening treatments while at the same time exhibiting the respective weak textures [9]. Table 1 shows a summary of the mechanical properties and stretch formability (IE) of several magnesium alloys.

Table 1. Summary of mechanical properties and Index Erichsen (IE) values of typical Mg alloy sheets compared to Al AA6xxx alloy sheet at RT (along the rolling direction; TYS: tensile yield strength, TE: total elongation).

Alloys (Wt%)	Processing Condition	TYS (MPa)	TE (%)	IE (mm)	References
AZ31	Hot rolling, annealing	166	23	2.6	[10]
AZ61	Extrusion, hot rolling, annealing	152	24	7.8	[11]
AZ80	Extrusion, hot rolling, annealing	187	24	3.7	[12]
Mg-3Al-1Zn-1Mn-0.5Ca	Twin-roll casting, hot rolling, annealing	219	16	8.0	[13]
AZ31	Hot rolling, annealing	179	22	3.4	[14]
WE43	Hot rolling, annealing	228	15	1.7	[15]
Mg-4.6Zn-0.6Ce-0.3La-0.2Nd	Hot rolling, T4	128	12	3.7	[16]
Mg-4.0Zn-0.3Y-0.3Ca	Twin-roll casting, hot rolling, annealing	176	26	7.6	[17]
AA6xxx	T4, naturally aged	180	27	9.1	[18]

Bian et al. (2020) [8] investigated the influence of different Ag contents on the mechanical–technological property profile of Mg-xAg-0.1Ca (0.3 ... 12 wt% Ag) alloys after extrusion and hot rolling. With the addition of 6 wt% Ag, the maximum values of IE 8.7 mm and yield strengths between 138 MPa (TD) and 182 MPa (RD) are achieved. The good formability at RT is attributed to the fine-grained microstructure accompanied by a weak TD-split texture, and the high strength to the dense distribution of fine AgMg4 particles. Jo et al. (2022) [9] showed an increased IE of 6.3 mm and yield strength values of 150 MPa after peak ageing of a ZAXM2100 magnesium alloy. The authors attribute the improved properties to a weakened texture with a basal pole split towards the RD and a broad angular distribution towards the TD as well as a fine-grained microstructure, which developed as a result of grain boundary pinning by Al2Ca. Weakened textures as a basis for the improved room temperature formability of several magnesium alloys are also reported by other research groups, for example, for AZ31 [19,20], ZMX21 [21] and

AZX612 [2]. Besides adding alloying elements, as mentioned above, appropriate processing technologies are also suitable for texture weakening. Han et al. 2023 [22] reported on the double-peak texture of AZ31 introduced to equal channel angular rolling combined with continuous bending and annealing, resulting in superior cold rolling formability.

Wang et al. (2021) [1] summarised in their review that RE- and Ca-containing magnesium alloys, which are subjected to extrusion or rolling–annealing procedures, reveal weak off-basal textures and homogeneous fine-grained microstructures. Microstructure and texture development are attributed to dynamic and static recrystallisation mechanisms associated with the change in the stacking fault energy and the activity of non-basal slip systems [1,13,17,23–25].

In the present work, tensile tests, and Erichsen tests are performed on and forming limit diagrams are created for the magnesium alloys AZ31 (Mg-3Al-1Zn) and ZAX210 (Mg-2Zn-1Al-0.3Ca), which were produced by the innovative twin-roll casting process and a following rolling process, to compare the effects of the different Mg alloys on the tensile properties and formability at room temperature. The insights gained from these investigations can contribute to the development of magnesium alloys with improved formability and enable their wider adoption in industries such as automotive, aerospace, and consumer electronics. The present work shows that sheet metal forming processes at low temperatures are possible with the ZAX210 alloy. In addition, a process whose industrial realisation has already been demonstrated is used via the production route of twin-roll casting combined with hot rolling. This means that the improved property profile is not just limited to laboratory-scale investigations.

2. Materials and Methods

2.1. Material

For the present investigations, magnesium sheets of the Mg-2Zn-1Al-0.3Ca (ZAX210) and Mg-3Al-1Zn (AZ31) alloys were employed. These sheets were produced through a combination of twin-roll casting and rolling processes, and a final annealing process. Twin-roll cast coils have a width of 730 mm. After trimming and hot rolling the final width of the coils was 650 mm. The resulting sheets had a uniform thickness of 1.5 mm. The chemical composition of the ZAX210 and the AZ31 strips, measured by means of radio spectral analysis, is provided in Table 2.

Table 2. Nominal chemical composition of ZAX210 and AZ31 alloys under investigation (wt.-%).

Alloy	Mg	Zn	Al	Ca	Mn	Cu	Fe	Ni	Si	Others
ZAX210	Bal.	2.29	0.92	<0.24	0.04	0.001	0.005	0.001	0.022	<0.020
AZ31	Bal.	0.85	2.92	0	0.32	0.001	0.005	0.001	0.021	<0.004

2.2. Experimental Procedure

The cast ingots were used as the initial material for the twin-roll casting process. The pilot plant for TRC at the Institute of Metal Forming, TU Bergakademie Freiberg, consists of a melting furnace, a casting channel and a casting nozzle. The twin-roll casting process was conducted at temperatures between 690 °C and 710 °C with casting speeds of 1.4 m/min. Further information can be found in [26]. The entire process takes place in an inert gas atmosphere.

The twin-roll cast sheets were homogenised at 430 °C (AZ31) or 380 °C (ZAX210) for 6 h. The hot rolling of the TRC sheets was performed in four passes from an initial thickness of 5.2 mm to a final thickness of 1.5 mm with an intermediate annealing stage at 430 °C (AZ31) or 380 °C (ZAX210) for 2 h. The four passes were from 5.2 mm to 3.6 mm to 2.7 mm and from 2.7 mm to 2 mm to 1.5 mm with a rolling speed of 100 m/min on a reverse rolling mill located in the Institute of Metal Forming at Technische Universität Bergakademie Freiberg, Germany. The rolling schedule was set with a strain of 0.28 and 0.37 in order to induce dynamic recrystallisation during hot rolling. After hot rolling, a final annealing was performed at 330 °C for 1 h. Samples for microstructural and texture characterisation were cut from the TRC, homogenised, hot rolled and finally annealed sheets. The choice of rolling and annealing parameters was based on [27] for AZ31 and [26] for ZAX210.

2.3. Material Characterisation

Samples for microstructure and texture analysis were metallographically prepared by grinding and polishing. Etching was carried out using a picric acid solution consisting of 70 mL of ethanol, 10 mL of distilled water, 10 mL of glacial acetic acid and 4.2 g of picric acid. Light microscopic images were taken using a Keyence VHX 6000 microscope at the Institute of Metal Forming, Freiberg, Germany. A scanning electron microscopic (SEM) evaluation was performed using a ZEISS GeminiSEM 450 device at the Institute of Metal Forming, Freiberg, Germany. Different detectors were used for SEM: angular selective backscatter (AsB), backscatter (BSE) and secondary electron detector (SE). Texture analysis was performed using an electron backscatter diffraction detector (EBSD). The accelerating voltage was between 15 and 20 kV. A step size of 0.65 µm was selected. The free MTEX MATLAB toolbox (version 5.8.1, MTEX, [28]) was used for the analysis of the EBSD data and the calculation of the pole figures. A microstructural analysis was carried out on longitudinal sections.

Tensile tests were performed to determine the tensile strength and total elongation of the materials according to DIN EN ISO 6892-1 [29]. Sample dimensions correspond to flat tensile samples, shape H, according to DIN 50125 [30]. The tests were carried out at room temperature, and the samples were loaded in the 0°, 45° and 90° directions with respect to the rolling direction. Five parallel tests were carried out in each case. An extensometer was used to measure the strain of a specimen under load. The results from the tensile tests provided valuable insights into the strength and ductility of the alloys.

To assess the stretch formability, Erichsen tests were conducted on the twin-roll cast, rolled and annealed strips according to DIN EN ISO 20482 [31]. The Erichsen test measures the ability of a sheet material to withstand plastic deformation without cracking or fracture. The average values of the Erichsen cup heights were recorded for both AZ31 and ZAX210 strips, providing quantitative data on their formability characteristics. By comparing the results, the relative formability of the alloys can be determined.

Furthermore, forming limit curves (FLCs) were created according to ISO 12004-2 [32] to analyse the formability of the alloys under different strain conditions. The Nakajima test, a widely used method for FLC determination, was employed. The test involves deforming specimens with varying notch widths using a hemispherical punch. A 1.00 ± 0.02 mm square grid was applied on the specimen surface via transfer printing for strain measurement which was conducted in situ with four cameras through the strain measuring system AutoGrid (ViALUX). The strains at which necking or fracture occurred provide crucial information about the formability limits of the materials. The FLCs obtained from the Nakajima tests allow for a comprehensive understanding of the sheet metal forming behaviour, considering the effects of stress condition, of both alloys.

3. Results

Characterisation of the Initial AZ31 and ZAX210 Strips

The inverse pole figure (IPF) maps, equivalent grain sizes and the backscattered electron images (BSE) of the twin-roll cast, rolled and annealed strips of AZ31 and ZAX210 are shown in Figures 1 and 2, respectively. Both alloys show approximately the same grain size: AZ31 has an average value of 7.9 µm, and ZAX210 has an average value of 8.9 µm. The AZ31 strips reveal finely distributed Al_8Mn_5 particles in α-Mg. The $Mg_{17}(Al,Zn)_{12}$ phases are dissolved according to [7]. The ZAX210 strips show Mg_2Ca-, $MgZn$- and $Ca_2Mg_6Zn_3$-phases in α-Mg, based on [33].

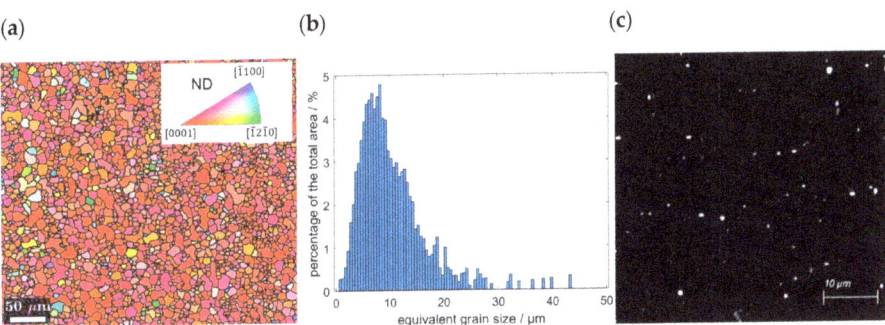

Figure 1. (**a**) Inverse pole figure (IPF) maps from longitudinal sections (TRC and rolling direction horizontal) of the twin-roll cast, rolled and annealed AZ31 strips, (**b**) equivalent grain sizes and (**c**) SE contrast.

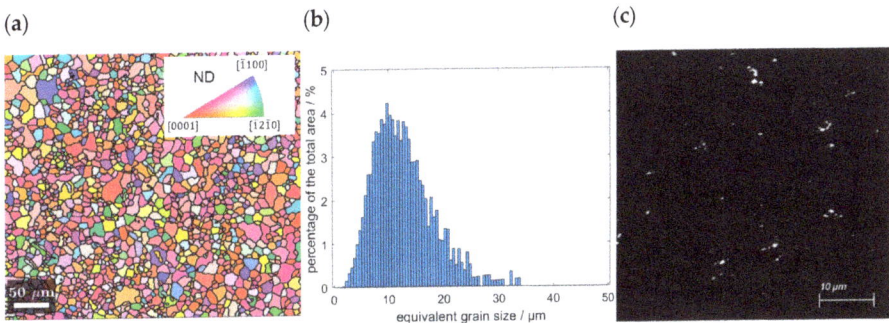

Figure 2. (**a**) Inverse pole figure (IPF) maps from longitudinal sections (TRC and rolling direction horizontal) of the twin-roll cast, rolled and annealed ZAX210 strips, (**b**) equivalent grain sizes and (**c**) SE contrast.

The influence of the alloying elements on the texture can be clearly seen in the IPFs and pole figures (Figure 3). Ca effectively weakens the intensity of the basal texture. AZ31 shows a strong and pronounced $\langle 0001 \rangle$ fibre parallel to the normal direction (ND) and a maximum intensity of and MRD of 10. Prismatic planes are randomly distributed parallel to the ND and weakly pronounced with 2.8 MRD. This texture is typical for rolled or extruded AZ31 strips [1,26]. The texture of ZAX210 differs significantly from that of AZ31. The basal pole figure (Figure 3b) shows the well-known split of the centre pole into two poles that tilt from the ND towards the RD. Ca can lead to this texture formation. Bian et al. (2020) [8] reported on RD split textures with double peaks after the hot rolling of a Ca-containing Mg-Ag magnesium alloy. Further studies also present weakened textures after the hot deformation of magnesium alloys after Ca addition [1,2,34]. Additionally, the maximum basal pole intensity is only 2.5 MRD. The prismatic planes show a weak preferred alignment in the TD with a maximum pole density of 1.7 MRD.

The mechanical properties of the twin-roll cast, rolled and annealed strips are shown in Table 3. The ultimate tensile strength (UTS) is not strongly influenced by a 0°, 45° or 90° direction. Only for the AZ31 alloy, there are some slightly higher values in 0° direction than in 90°.

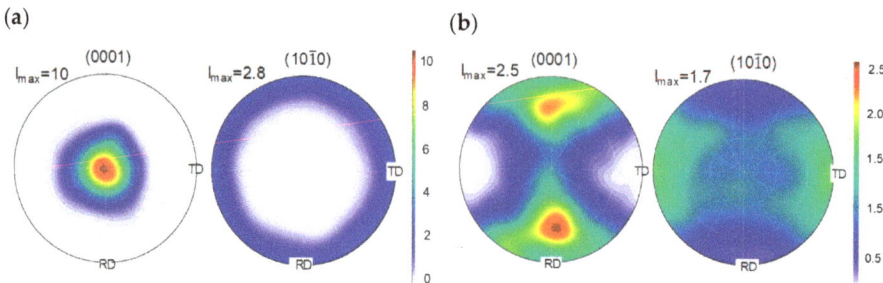

Figure 3. Pole figures of the twin-roll cast, rolled and annealed Mg strips. (**a**) AZ31, (**b**) ZAX210 (intensities in multiples of random distribution—MRD).

Table 3. Mechanical properties and IE of the twin-roll cast, rolled and annealed Mg strips.

Strips	Direction	TYS (MPa)	UTS (MPa)	TE (%)	UTS/YS	I.E.
AZ31	RD	198 ± 5	280 ± 6	24 ± 2	1.4	
	45°	195 ± 6	276 ± 6	22 ± 2	1.4	4.9 ± 0.2
	TD	189 ± 5	273 ± 5	18 ± 1	1.4	
ZAX210	RD	159 ± 3	248 ± 4	30 ± 2	1.6	
	45°	145 ± 3	238 ± 3	32 ± 2	1.6	7.1 ± 0.3
	TD	136 ± 3	242 ± 4	28 ± 3	1.8	

The lowest total elongation (TE) was found at 90°, and the largest one was found at 0° for both alloys, in which the difference in AZ31 strips is much higher than that in the ZAX210 strips.

At room temperature, the ZAX210 alloy exhibits medium strengths with tensile strengths of about 240 MPa. The 0.2% proof stress (TYS) varies between 136 MPa (TD, 90°) and 159 MPa (RD, 0°) depending on the sampling direction. The Excellent total elongation of 28% (TD, 90°) to 32% (45°) are achieved. The deviation between the mechanical properties in and transverse to the rolling direction is a maximum of 15%. A comparison with AZ31 shows higher strengths than ZAX210 due to the high aluminium content. However, the directional dependence of the mechanical properties is more pronounced, due to the strong basal texture formed during hot rolling. For example, transverse to the rolling direction, only about 50% of the total elongation can be achieved in the rolling direction.

In comparison with the literature on rare-earth-containing magnesium alloy ZE10, which exhibits a weak basal texture with intensity maxima along the rolling direction, pole broadening in the transverse direction in the rolling direction and finally annealed state shows that similar mechanical properties are achieved for ZAX210. Due to the basal pole broadening in TD for ZE10, higher total elongations in the 90°-direction are achieved [26]. These results are in agreement with the literature [1,24].

To investigate the impact on stretch formability at room temperature, Erichsen tests were performed on both alloys. An average Erichsen Index (I.E.) of 4.9 mm and 7.1 mm were recorded for twin-roll cast, rolled and annealed AZ31 and ZAX210 strips, respectively. The significant high formability of the ZAX210 strip is closely related to the texture softening originating from the weakening of basal texture intensity and the splitting of the basal pole towards the RD. Generally, the stretch formability is mainly dependent on the microstructure and texture of the sheets and the associated forming mechanisms. Particle-stimulated nucleation (PSN), recrystallisation at deformation and shear bands, changes in grain boundary mobility brought on by secondary phases, and recrystallisation at double and compression twins are possible mechanisms for the weakening of textures [1,2,8,21].

As written in Kittner et al. (2022) [33] recrystallisation due to PSN can be excluded in the current ZAX210 strip. Rather, this strongly weakened texture, in which the c-axis of the crystallites is tilted from the normal direction (ND) to the RD-direction primarily, is caused by continuous or twin-induced dynamic recrystallisation at double and compression twins. Additionally, it is possible that Ca, which segregates at grain boundaries, limits their mobility. As a result, the intensity of the resulting texture is affected, and the preferential growth of grains is inhibited [33]. In Wang et al. (2021) [1], the weak off-basal texture after extrusion or rolling-annealing procedures is attributed to a homogeneous fine-grained microstructure as a result of dynamic and/or static recrystallisation associated with the activity of the non-basal slip systems and the chance of stacking fault energy.

Figure 4 plots the Erichsen Index and tensile yield strength values of AZ31 and ZAX210 of this study at room temperature compared to other magnesium alloys from the literature as well as the AA6xxx aluminium alloy sheet. Compared to the commercial AZ31, AZ80 and WE43 alloys, the twin-roll cast, hot rolled and finally annealed AZ31 and ZAX210 alloys show improved room temperature formability with comparable tensile yield strength (except in relation to WE43). Compared to other newly developed magnesium alloys with Ca addition, the ZAX210 alloy exhibits a slightly lower room temperature formability but has excellent elongation at fracture. The outstanding combination of strength, ductility and formability of the ZAX210 magnesium alloy, which is comparable to those of AA6xxx aluminium alloy sheets, leads to an increased potential of those magnesium alloys for a wide range of applications, especially in the field of room-temperature forming.

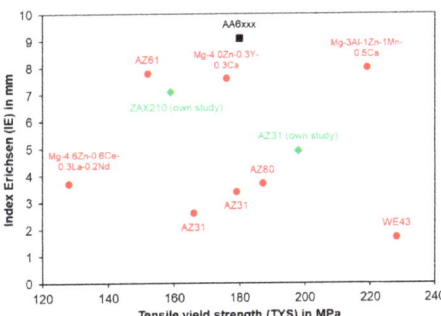

Figure 4. Index Erichsen (IE) values versus tensile yield strength (TYS) of various magnesium alloys (red) from the literature (according to Table 1) compared to an aluminium alloy (black) as well as AZ31 and ZAX210 from this study (green), adapted from [10–18].

For the further analysis of cold forming, forming limit strains are determined at room temperature using the Nakajima test. It is important to note that the forming operation along with the left side of the FLC are based on the strain condition that allows the material to contract along the minor strain axis but with little to no reduction in thickness. In contrast, on the right side of the wide FLC, the forming operation requires a positive strain in all the sheet plane directions, and the material flow is realised by thinning the sheet. The overall minimum depends on the plane strain condition in V-shaped curves. Such a result can be seen for the ZAX210 and AZ31 sheets in Figure 5. As expected, the forming limit value for ZAX210 sheet is higher than for AZ31 sheet. Compared to the above forming operation, the ZAX210 strains on the left side of the FLC are higher than those on the right side. For forming operations with a deep-drawing component as well as uniaxial tension, higher forming changes are therefore possible at room temperature. In the range of negative minor strain, the major strain is $\varphi_1 = 0.35$. The stretch-forming capacity and the forming capacity in the plane strain state are at a comparable level with $\varphi_1 = 0.2$.

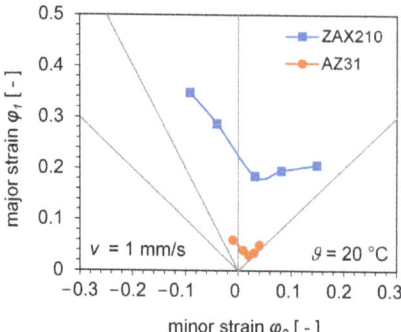

Figure 5. Forming limit curves of the twin-roll cast, rolled and annealed ZAX210 and AZ31 strips at room temperature.

In comparison with the literature values, the ZAX210 finished strip exhibits a comparable, partly higher forming capacity than strips of a ZE10 alloy [35]. Due to their chemical composition—ZE10 as a result of the addition of rare earth elements and ZAX210 as a result of the addition of Ca—both alloys have weakened textures after rolling, which are generally accompanied by improved formability.

4. Conclusions

The investigation focused on the characterisation of twin-roll cast, hot rolled and finally annealed magnesium sheets of the ZAX210 (Mg-2Zn-1Al-0.3Ca) alloy and its comparison with the AZ31 (Mg-3Al-1Zn) alloy.

i. Both alloys exhibited similar grain sizes, with AZ31 having an average value of 7.9 µm and ZAX210 having an average value of 8.9 µm. The microstructure of the AZ31 strips showed finely distributed Al_8Mn_5 particles in α-Mg, while the ZAX210 strips exhibited Mg_2Ca, $MgZn$, and $Ca_2Mg_6Zn_3$ phases in α-Mg.

ii. The influence of alloying elements on texture was evident in the IPFs and pole figures. The presence of calcium effectively weakened the intensity of the basal texture in ZAX210, whereas AZ31 exhibited a strong basal texture with a pronounced $\langle 0001 \rangle$ fibre parallel to the normal direction (ND). The mechanical properties, including tensile strength and total elongation, were evaluated for both alloys. ZAX210 demonstrated medium strengths with tensile strengths of about 240 MPa and excellent total elongation ranging from 28% to 32%, depending on the sampling direction. A comparison with AZ31 revealed higher strengths in AZ31 due to its higher aluminium content, but also a stronger directional dependence in mechanical properties attributed to its strong basal texture formed during hot rolling.

iii. The formability at room temperature of the alloys was assessed through Erichsen tests, with ZAX210 strips demonstrating significantly higher formability compared to AZ31. The texture softening in ZAX210, resulting from the weakening of basal texture intensity and the splitting of basal poles towards the rolling direction (RD), contributed to its enhanced formability. Recrystallisation at double and compression twins, along with the potential grain boundary segregation of calcium, were identified as mechanisms for the weakened texture in ZAX210 [33]. The ZAX210 alloy offers an outstanding combination of strength, ductility and formability, which is comparable to those of AA6xxx aluminium alloy sheets.

iv. Forming limit strains were determined using the Nakajima test, indicating that ZAX210 exhibited higher forming limit values compared to AZ31. The ZAX210 strip demonstrated higher forming changes on the left side of the forming limit curve (FLC), allowing for deeper forming operations at room temperature. The forming capacity of ZAX210 is comparable to a ZE10 alloy, which contains rare earth elements, due to the weakened textures achieved after rolling.

In summary, the study revealed that the ZAX210 alloy, with the addition of calcium, exhibited favourable mechanical properties, enhanced formability, and weakened textures, making it a promising material for cold forming applications. In direct comparison to the well-established AZ31 alloy, this new alloy shows much improved formability what will make it a high-potential candidate for future cold forming operations. However, despite the mentioned advances of the ZAX210 magnesium alloy, further research in the field of a simultaneous enhancement in strength, ductility and stretch formability at room temperature is still required. Age or bake hardening procedures are planned for the twin-roll cast, hot rolled and annealed ZAX210 sheets with the aim to increase strength by precipitation hardening.

Author Contributions: Conceptualisation, M.U.; methodology, M.U.; software, M.U. and K.K.; validation, M.U. and K.K.; investigation, M.U.; data curation, K.K.; writing—original draft, M.U.; writing—review and editing, K.K. and U.P.; visualisation, M.U.; supervision, U.P.; project administration, M.U. and K.K.; funding acquisition, U.P. All authors have read and agreed to the published version of the manuscript.

Funding: This study arises within the research project CLEAN-Mag, which is funded by the Federal Ministry for Economic Affairs and Climate Action (BMWK) as part of the Lightweight Construction Technology Transfer Programme (TTP Leichtbau) and managed by Project Management Jülich (PTJ). The authors would like to thank the funding organisations for their support.

Data Availability Statement: The data are contained within the document otherwise can be shared upon request to the corresponding author.

Conflicts of Interest: The authors declare no conflict of interest.

References

1. Wang, Q.; Jiang, B.; Chen, D.; Jin, Z.; Zhao, L.; Yang, Q.; Huang, G.; Pan, F. Strategies for enhancing the room-temperature stretch formability of magnesium alloy sheets: A review. *J. Mater. Sci.* **2021**, *56*, 12965–12998. [CrossRef]
2. Bian, M.; Huang, X.; Chino, Y. Substantial improvement in cold formability of concentrated Mg-Al-Zn-Ca alloy sheet by high temperature final rolling. *Acta Mater.* **2021**, *220*, 117328. [CrossRef]
3. Krajňák, T.; Minárik, P.; Stráský, J.; Máthis, K.; Janeček, M. Mechanical properties of ultrane-grained AX41 magnesium alloy at room and elevated temperatures. *Mater. Sci. Eng. A Struct. Mater. Prop. Microstruct. Process.* **2018**, *731*, 438–445. [CrossRef]
4. Mendis, C.; Bae, J.; Kim, N.; Hono, K. Microstructures and tensile properties of a twin roll cast and heat-treated mg-2.4Zn-0.1Ag-0.1Ca-0.1Zr alloy. *Scr. Mater.* **2011**, *64*, 335–338. [CrossRef]
5. Hänzi, A.C.; Sologubenko, A.S.; Gunde, P.; Schinhammer, M.; Uggowitzer, P.J. Design considerations for achieving simultaneously high-strength and highly ductile magnesium alloys. *Philos. Mag. Lett.* **2012**, *92*, 417–427. [CrossRef]
6. Chai, Y.; Jiang, B.; Song, J.; Wang, Q.; Gao, H.; Liu, B.; Huang, G.; Zhang, D.; Pan, F. Improvement of mechanical properties and reduction of yield asymmetry of extruded mg-sn-zn alloy trough Ca addition. *J. Alloys Compd.* **2019**, *782*, 1076–1086. [CrossRef]
7. Kammer, C. *Magnesium Taschenbuch*; Aluminium-Verlag: Düsseldorf, Germany, 2000.
8. Bian, M.; Huang, X.; Chino, Y. A combined experimental and numerical study on room temperature formable magnesium-silver-calcium alloys. *J. Alloys Compd.* **2020**, *834*, 155017. [CrossRef]
9. Jo, S.; Bohlen, J.; Kurz, G. Individual Contribution of Zn and Ca on Age-Hardenability and Formability of Zn-Based Magnesium Alloy Sheet. *Materials* **2022**, *15*, 5239. [CrossRef]
10. Huang, X.; Suzuki, K.; Watazu, A.; Shigematsu, I.; Saito, N. Improvement of formability of Mg-Al-Zn alloy sheet at low temperatures using differential speed rolling. *J. Alloys Compd.* **2009**, *470*, 263–268. [CrossRef]
11. Huang, X.; Suzuki, K.; Chino, Y.; Mabuchi, M. Texture and strecht formability of AZ61 and AM60 magnesium alloy sheets processed by high-temperature rolling. *J. Alloys Compd.* **2015**, *632*, 94–102. [CrossRef]
12. Huang, X.; Suzuki, K.; Saito, N. Microstructure and mechanical properties of AZ80 magnesium alloy sheet processed by differential speed rolling. *Mater. Sci. Eng. A* **2009**, *508*, 226–233. [CrossRef]
13. Trang, T.T.T.; Zhang, J.H.; Kim, J.H.; Zargaran, A.; Hwang, J.H.; Suh, B.-C.; Kim, N.J. Designing a magnesium alloy with hight strength and high formability. *Nat. Commun.* **2018**, *9*, 2522. [CrossRef] [PubMed]
14. Chino, Y.; Mabuchi, M. Enhanced strech formability of Mg-Al-Zn alloy sheets rolled at high temperatuer (723 K). *Scirpta Mater.* **2009**, *60*, 447–450. [CrossRef]
15. Suh, B.-C.; Shim, M.-S.; Shin, K.; Kim, N.J. Current issues in magenium sheet alloys: Where do we go from here? *Scr. Mater.* **2014**, *84–85*, 1–6. [CrossRef]

16. Wu, Z.; Ahmad, R.; Yin, B.; Sandlöbes, S.; Curtin, W.A. Mechanistic origin and prediction of enhanced ductility in magnesium alloys. *Science* **2018**, *359*, 447–452. [CrossRef]
17. Park, S.; Jung, H.C.; Shin, K.S. Deformation Behavior of Twin Roll Cast Mg-Zn-X-Ca Alloys for Enhanced Room-Temperature Formability. *Mater. Sci. Eng. A* **2017**, *679*, 329–339. [CrossRef]
18. Liu, H.; Zhao, G.; Liu, C.-M.; Zuo, L. Effects of different tempers on precipitation hardening of 6000 series aluminium alloys. *Trans. Nonferrous Met. Soc. China* **2007**, *17*, 122–127. [CrossRef]
19. Zhang, H.; Ren, S.; Li, X.; Wang, L.; Fan, J.; Chen, S.; Zhu, L.; Meng, F.; Tong, Y.; Roven, H.J.; et al. Dramatically enhanced stamping formability of Mg-3Al-1Zn alloy by weakening (0001) basal texture. *J. Mater. Res. Technol.* **2020**, *9*, 14742–14753. [CrossRef]
20. Chaudry, U.M.; Hamad, K.; Kim, J. A Further Improvement in the Room-Temperature Formability of Magnesium Alloy Sheets by Pre-Stretching. *Materials* **2020**, *13*, 2633. [CrossRef]
21. Bian, M.; Huang, X.; Chino, Y. Towards Improving Cold Formability of a Concentrated Mg-Al-Zn-Ca Alloy Sheet. In *Magnesium Technology*; Springer International Publishing: Cham, Switzerland, 2022; pp. 227–231. [CrossRef]
22. Han, X.-Z.; Hu, L.; Jia, D.-Y.; Chen, J.-M.; Zhou, T.; Jiang, S.-Y.; Tian, Z. Role of unusual double-peak texture in significantly enhancing cold rolling formability of AZ31 magnesium alloy sheet. *Trans. Nonferrous Met. Soc. China* **2023**, *33*, 2351–2364. [CrossRef]
23. Zhang, B.; Wang, Y.; Geng, L.; Lu, C. Effects of calcium on texture and mechanical properties of hot-extruded Mg-Zn-Ca alloys. *Mater. Sci. Eng. A Struct. Mater. Prop. Microstruct. Process.* **2012**, *539*, 56–60. [CrossRef]
24. Chino, Y.; Ueda, T.; Otomatsu, Y.; Sassa, K.; Huang, X.; Suzuki, K.; Mabuchi, M. Effects of Ca on tensile properties and stretch formability at room temperature in Mg-Zn and Mg-Al alloys. *Mater. Trans.* **2011**, *52*, 1477–1482. [CrossRef]
25. Griffiths, D. Explaining texture weakening and improved formability in magnesium rare earth alloys. *Mater. Sci. Technol.* **2015**, *31*, 10–24. [CrossRef]
26. Ullmann, M.; Kittner, K.; Henseler, T.; Stöcker, A.; Prahl, U.; Kawalla, R. Development of new alloy systems and innovative processing technologies for the production of magnesium flat products with excellent property profile. *Procedia Manuf.* **2019**, *27*, 203–208. [CrossRef]
27. Neh, K.; Ullmann, M.; Oswald, M.; Berge, F.; Kawalla, R. Twin ro ll casting and strip rolling of several magnesium alloys. *Mater. Today Proc.* **2015**, *2*, S45–S52. [CrossRef]
28. Bachmann, F.; Hielscher, R.; Schaeben, H. Texture Analysis with MTEX–Free and Open Source Software Toolbox. *Solid State Phenom.* **2010**, *160*, 63–68. [CrossRef]
29. *DIN EN ISO 6892-1*; Metallische Werkstoffe—Zugversuch—Teil1: Prüfverfahren bei Raumtemperatur. Deutsches Institut für Normung: Berlin, Germany, 2019.
30. *DIN 50125*; Prüfung metallischer Werkstoffe—Zugproben. Deutsches Institut für Normung: Berlin, Germany, 2022.
31. *DIN EN ISO 20482*; Metallische Werkstoffe—Bleche und Bänder—Tiefungsversuch nach Erichsen. Deutsches Institut für Normung: Berlin, Germany, 2013.
32. *ISO 12004-2*; Metallischer Werkstoffe—Bestimmung der Grenzformänderungskurven für Bleche und Bänder—Bestimmung von Grenzformänderungskurven im Labor. Deutsches Institut für Normung: Berlin, Germany, 2021.
33. Kittner, K.; Ullmann, M.; Prahl, U. Microstructural andTextural Investigation of an Mg-Zn-Al-Ca Alloy after Hot PlaneStrain Compression. *Materials* **2022**, *15*, 7499. [CrossRef]
34. Nienaber, M.; Bohlen, J.; Victoria-Hernández, J.; Yi, S.; Kainer, K.U.; Letzig, D. Cold Formabiliy of Extruded Magnesium Bands. In *Magnesium Technology*; Springer International Publishing: Cham, Switzerland, 2020; pp. 329–334. [CrossRef]
35. Stutz, L. Das Umformverhalten von Magnesiumblechen der Legierungen AZ31 und ZE10. Ph.D. Dissertation, Technische Universität Berlin, Berlin, Germany, 2015.

Disclaimer/Publisher's Note: The statements, opinions and data contained in all publications are solely those of the individual author(s) and contributor(s) and not of MDPI and/or the editor(s). MDPI and/or the editor(s) disclaim responsibility for any injury to people or property resulting from any ideas, methods, instructions or products referred to in the content.

Article

BCC-Based Mg–Li Alloy with Nano-Precipitated MgZn$_2$ Phase Prepared by Multidirectional Cryogenic Rolling

Qing Ji, Xiaochun Ma, Ruizhi Wu *, Siyuan Jin, Jinghuai Zhang and Legan Hou

Key Laboratory of Superlight Materials & Surface Technology (Ministry of Education), Harbin Engineering University, Harbin 150001, China
* Correspondence: rzwu@hrbeu.edu.cn

Abstract: In this study, we deformed the single β phase Mg–Li alloy, Mg–16Li–4Zn–1Er (LZE1641), with conventional rolling (R) and multi-directional rolling (MDR), both at cryogenic temperature. Results showed that the nano-precipitation phase MgZn$_2$ appeared in the alloy after MDR, but this phenomenon was not present in the alloy after R. The finite element simulation result showed that the different deformation modes changed the stress distribution inside the alloy, which affected the microstructures and the motion law of the solute atoms. The high-density and dispersively distributed MgZn$_2$ particles with a size of about 35 nm were able to significantly inhibit the grain boundary migration. They further hindered the dislocation movement and consolidated the dislocation strengthening and fine-grain strengthening effects. Compared with the compressive strength after R (273 MPa), the alloy compressive strength was improved by 21% after MDR (331 MPa). After 100 °C compression, the MgZn$_2$ remained stable.

Keywords: bcc Mg–Li alloy; cryogenic; multi-directional rolling; nano–grains; dislocation; MgZn$_2$ phase

Citation: Ji, Q.; Ma, X.; Wu, R.; Jin, S.; Zhang, J.; Hou, L. BCC-Based Mg–Li Alloy with Nano-Precipitated MgZn$_2$ Phase Prepared by Multidirectional Cryogenic Rolling. *Metals* **2022**, *12*, 2114. https://doi.org/10.3390/met12122114

Academic Editor: Koh-ichi Sugimoto

Received: 12 November 2022
Accepted: 6 December 2022
Published: 8 December 2022

Publisher's Note: MDPI stays neutral with regard to jurisdictional claims in published maps and institutional affiliations.

Copyright: © 2022 by the authors. Licensee MDPI, Basel, Switzerland. This article is an open access article distributed under the terms and conditions of the Creative Commons Attribution (CC BY) license (https://creativecommons.org/licenses/by/4.0/).

1. Introduction

Magnesium (Mg) alloys not only have the advantages of low density, high specific strength, and high specific stiffness, but also have excellent damping and electromagnetic shielding properties [1,2]. These properties enable Mg alloys to adapt well to the high demands of aerospace, military equipment, medical equipment, and 3C electronics. However, in the production process, Mg alloys have a strong basal texture and are difficult to deform due to the close-packed hexagonal (hcp) lattice structure of Mg [3–5], which weakens their advantages when competing with other lightweight alloys.

The addition of Li into Mg alloys gradually changes the hcp structure to body-centered cubic (bcc) structure. Depending on the Li content, the lattice type of magnesium alloys present three states: α phase (<5.7 wt.%), α + β phase (5.7–10.3 wt.%), and β phase (>10.3 wt.%) [6–8]. Moreover, the density of Li is 0.53 g/cm^3, which can further reduce the density of the Mg alloy. To cope with the strict demands for lightweight alloys, 14 wt.% or more Li is added to the Mg alloy. However, the addition of Li introduces a new problem, namely that it is difficult to improve the strength of the alloy beyond 200 MPa. The increase of active slip systems in bcc lead to the easy deformation of the alloy and the easy activation of dislocations [9]. Therefore, improving the strength of single-β-phase Mg–Li alloys has become a particularly urgent focus.

To overcome such problems, research has mainly focused on plastic deformation [10,11]. Plastic deformation processes such as extrusion, rolling, and torsion are effective means to improve the strength of β-phase magnesium–lithium alloys. The bcc structure has high stacking fault energy, wide spreading dislocations, and low critical shear stress (CRSS). These characteristics make the dislocation prone to slip [12]. However, to counteract the stress-concentration state, dynamic recrystallization (DRX) prematurely enters the stage of grain growth after nucleation, which causes dislocations to be prematurely eliminated during the accumulation process. Thus, the effect of work-hardening and fine-grain strengthening cannot continue to proliferate with the amount of deformation.

Changing parameters such as temperature and deformation can effectively improve the strength of β-phase Mg–Li alloys by regulating DRX and dislocations [13–16]. Furthermore, solid solution strengthening and second-phase strengthening play important roles in Mg–Li alloys [17,18]. In addition to the primary second phase in the as-cast state, the dispersive reinforced phase dynamically precipitated during the deformation process also enables the optimization of the mechanical properties [19].

The present study compared the microstructures and mechanical properties at different temperature and deformation modes (R and MDR). High-density, dispersively distributed, nano-scale $MgZn_2$ particles and refined grains were found in the specimen deformed by cryogenic MDR. The relative strengthening mechanisms in the specimen are discussed.

2. Material and Methods

2.1. Preparation and Processing of Raw Materials

Mg–16Li–4Zn–1Er (wt.%) alloy was prepared in a vacuum medium-frequency induction furnace under the protection of Ar atmosphere. The ingots for the alloy were pure metals of Mg (>99.9 wt.%), Li (>99.9 wt.%), Zn (>99.9 wt.%), and a master alloy of Mg–20 wt.%Er. Then, the melt was poured into a permanent mold with dimensions of 120 mm × 110 mm × 40 mm. The 20 mm × 20 mm × 20 mm blocks for rolling were cut from the as-cast alloy. The rolling modes were R and MDR. The intended reduction between each pass was set at 0.8 mm. The specific steps of the processing are shown in Figure 1 (RD, rolling direction; TD, transverse direction; ND, normal direction). The total reduction of R was 60%, while thr A-side was always maintained as an RD–TD surface. The formula for the equivalent strain law of rolling process is as follows [20]:

$$\varepsilon = \frac{2}{\sqrt{3}} \left| \ln\left(\frac{h_0}{h}\right) \right|$$

In this experiment, the equivalent strain of MD rolling was stipulated to be in line with R, that is, $\varepsilon_{MDR} = \varepsilon_{R60\%} = 1.05804$. In MD rolling, A-side and B-side act as the RD–TD surface and successively consume half of the equivalent strain ($\varepsilon_{A-MDR} = \varepsilon_{B-MDR} = 1/2\varepsilon_{MDR}$). The details are shown in Figure 1. Importantly, after many experimental operations, it was shown that in the actual rolling process, due to the lateral flow of metal, when the thickness of sample was reduced from 20 mm to 12.6 mm, the width expanded from 20 mm to 23 mm.

For cryogenic rolling, the specimens were steeped in liquid nitrogen for 10 min before rolling. Similarly, they were placed back into liquid nitrogen for 5 min between each lane.

2.2. Microstructural Characterization

Transmission electron microscopy (TEM) and X-ray diffraction (XRD) were used for the analysis of the microstructures. The average grain size and second-phase size were determined by the mean linear intercept method.

2.3. Compression Test

The gleeble experiments were carried out at 25 °C and 100 °C. The size of the sample was ϕ8 mm × 12 mm. The position of the compression sample in the rolled samples is shown in Figure 1. The strain rate was 1.0×10^{-2} s^{-1}, and the engineering strain was 70%. The mean value for each state was determined by the five compressed samples.

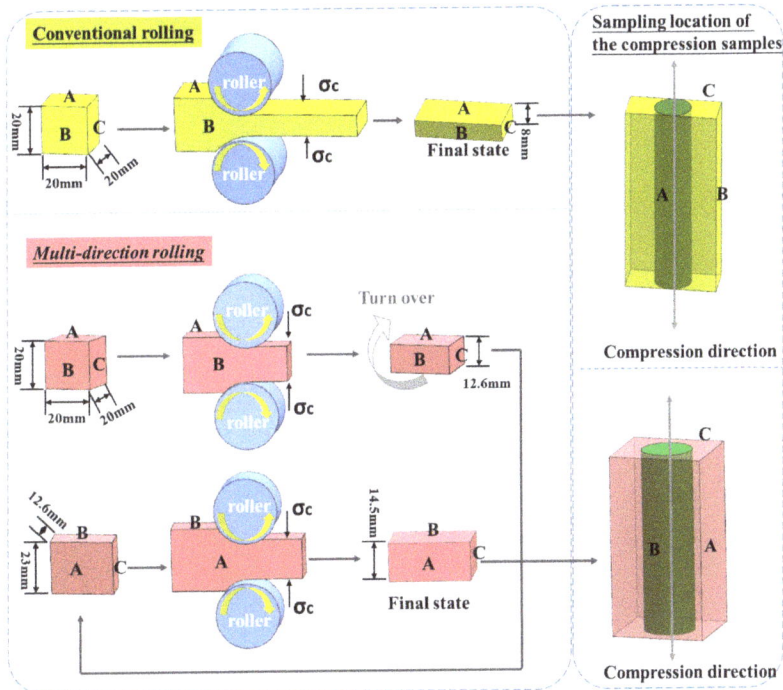

Figure 1. Schematic diagram of rolling and the positions of compression samples in as-rolled samples.

3. Results

3.1. Microstructures

The TEM bright-field images of the areas without the second-phase fragments were analyzed as shown in Figure 2. When the alloy grains were rolled in multiple directions at cryogenic temperature, a large range of uniformly distributed nano-grains appeared, with a size of about 56 nm. When a diffraction spot image was collected in any area of the nanocrystal, it contained all ring-shaped spots.

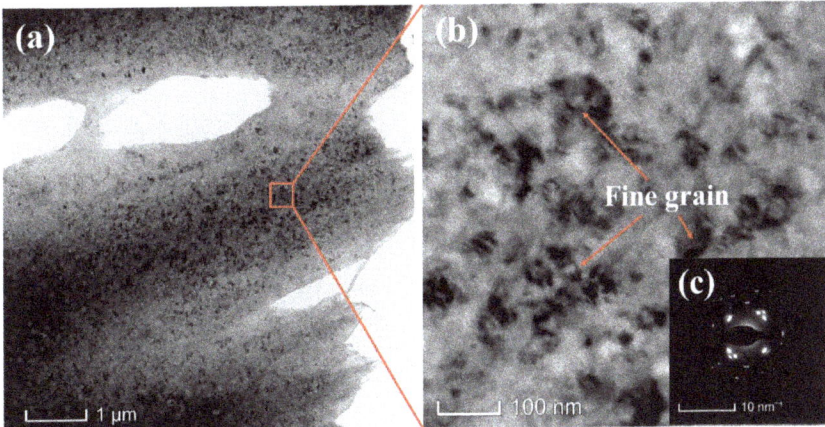

Figure 2. (a,b) Fine grains in bright-field TEM images of cryogenic MD-rolled alloy; (c) diffraction spot image.

It is worth noting that the nano-MgZn$_2$ phase was detected in the matrix of the cryogenic MD-rolled alloy. Figure 3a shows the bright-field TEM image area that was selected from the cryogenic MD rolled alloy for the measurement of diffraction spots. Figure 3b is a composite spot image of the β-Li and MgZn$_2$ phases. Figure 3c,d show dark-field TEM images of MgZn$_2$ under different crystal planes. In the LZE1641 alloy system, the primary second phase is a micron-sized second-phase particle, and there is no nano-sized second phase. Based on the results of our previous study [13], we inferred that this nanometer-sized densely distributed second phase was precipitated during the deformation process; that is, the morphology of small particles and diffraction spots in the dark field can be identified as the MgZn$_2$ precipitated phase. According to the present results, the precipitated phase was densely and uniformly distributed in the matrix with a size of about 35 nm. Figure 4 shows the XRD image of MD rolling at cryogenic temperature. In XRD, the peak of the MgZn$_2$ phase appeared on the (301) crystal plane.

The bright-field TEM images of the cross-section after compression at 100 °C are shown in Figure 5. After the hot compression, the grain size of the cryogenic rolled alloy was about 2~4 μm while the grain size of the cryogenic MD-rolled alloy was 1 μm. In addition, a large number of MgZn$_2$ particles still remained in the matrix. The TEM image of the nano-precipitated MgZn$_2$ phase area and the detection result of the element distribution are shown in Figure 6. We found that the average size of the nano-phase MgZn$_2$ phase was about 40 nm, which is the same as the size before compression. The distribution of Zn was consistent with the distribution of the nano-precipitated phase. This demonstrates that the MgZn$_2$ phase could not be fragmented after hot compression and is conducive to hindering the migration of grain boundaries during said compression [21].

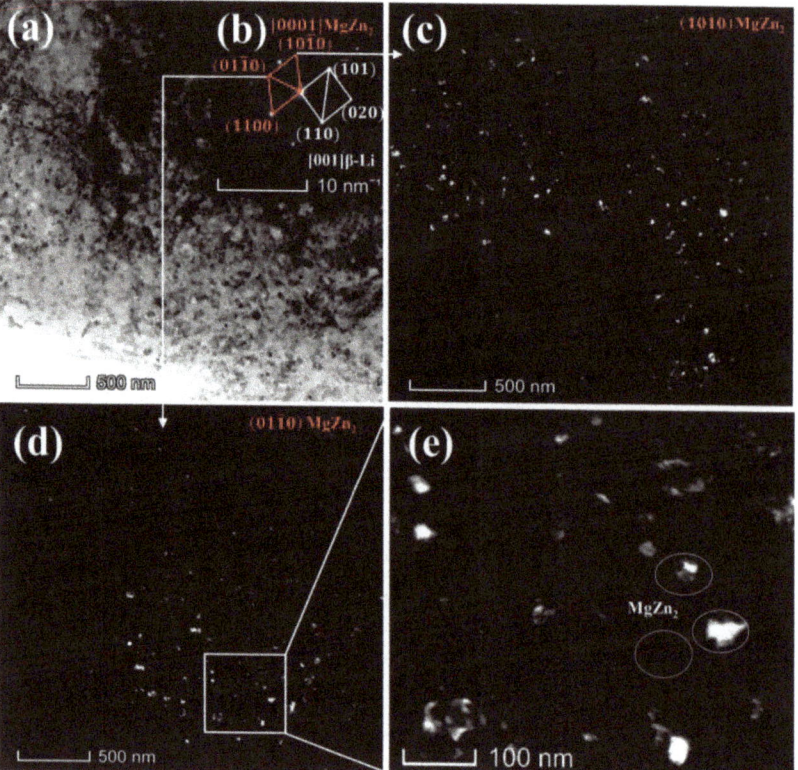

Figure 3. (a) Bright-field TEM images of cryogenic MD-rolled alloy; (b) diffraction spot image of (a); (c–e) dark-field TEM images of nano-MgZn$_2$ phases.

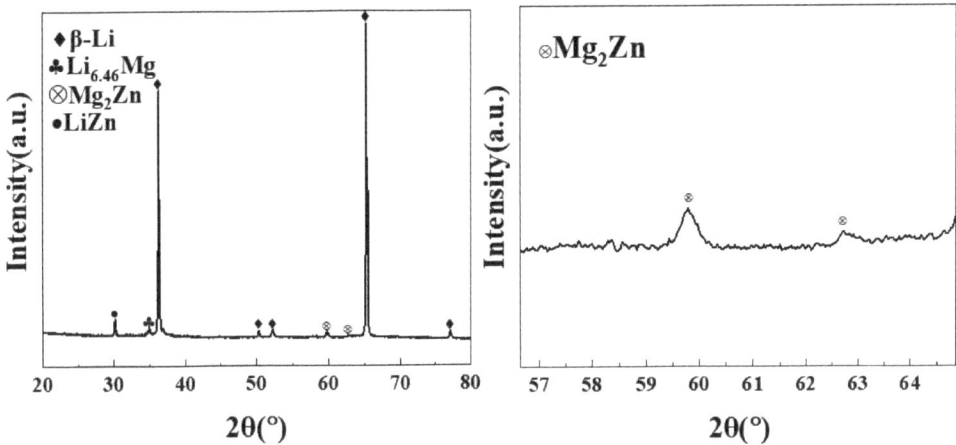

Figure 4. XRD images of cryogenic MD-rolled alloy.

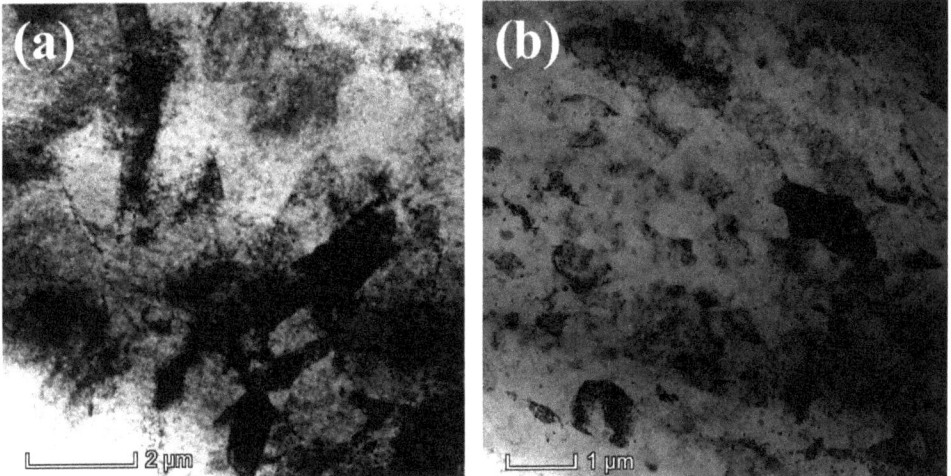

Figure 5. Bright-field TEM images of cross-section morphologies after 100 °C compression, (**a**) cryogenic conventionally rolled alloy; (**b**) cryogenic MD-rolled alloy.

As shown in Figure 7, while the dislocations are arranged in a disordered manner at the grain boundary, the nano-precipitation particles further aggravated the proliferation and packing of the dislocations. Multiple dislocation lines are intertwined and entangled with each other, providing an effect for the work-hardening of the alloy. Based on these results, MgZn$_2$ only existed in the MD-rolled alloy at the cryogenic temperature. This indicates that MgZn$_2$ is a stress-induced nano-precipitation phase. The type, structure, and shape of the second phase depend on the elastic strain energy inside the alloy [22]. To eliminate the concentrated stress level of the cryogenic MD-rolled alloy, spherical nano-MgZn$_2$ with hcp structure is precipitated. This nano-precipitated phase is difficult for dislocations to cut and can effectively pin their movement. This promotes the intertwining phenomenon of dislocations [23]. In the cryogenic environment, dislocations cannot be released immediately, which further enhances the number of dislocations and the degree of dislocation packing in the matrix. Thus, dislocation strengthening was further optimized.

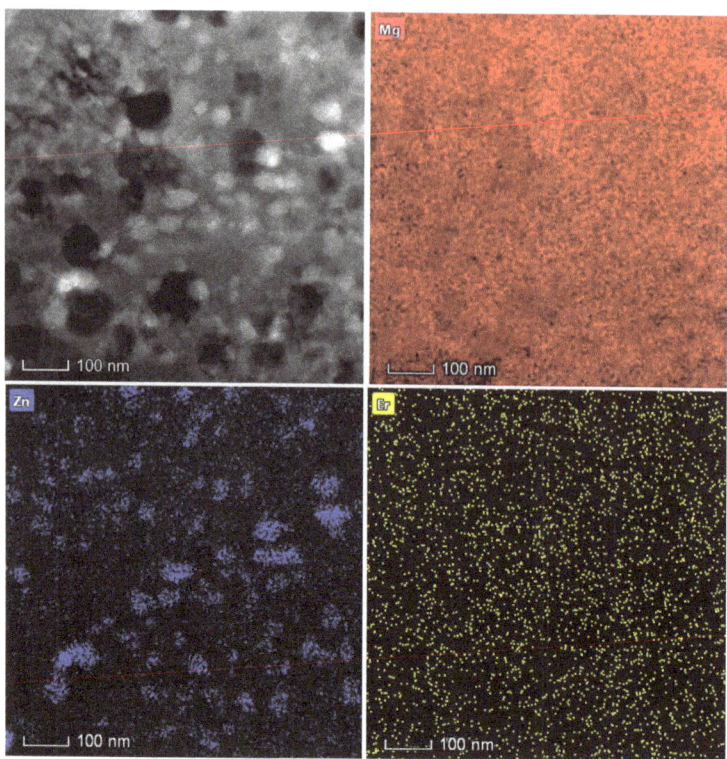

Figure 6. Mappings of nano-MgZn$_2$ phases in cryogenic MD-rolled alloy after 100 °C compression.

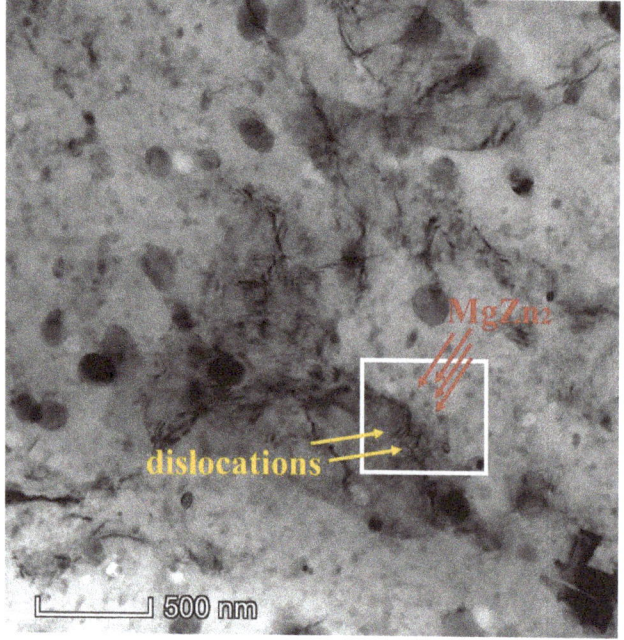

Figure 7. Bright-field TEM images for the interaction between nano-MgZn$_2$ phases and dislocations.

3.2. Mechanical Properties

The true stress–strain curves of cryogenic rolled alloys compressed at 25 °C (room temperature) and 100 °C are shown as Figure 8. The compressive strength of the MDR alloy was 331 MPa, which was 21% higher than that of the conventional rolled alloy (273 MPa). After compression at 100 °C, the compressive strength of the conventional rolled alloy was 192 MPa, while the MDR alloy could still maintain its compressive strength of 238 MPa.

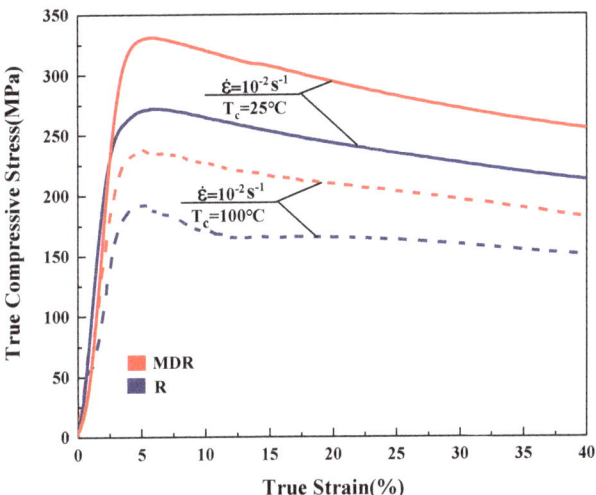

Figure 8. Compressive true stress–true strain curves of cryogenic rolled alloys.

4. Discussion

4.1. Precipitation of the MgZn$_2$ Phase

In our previous experimental results, we proved that low temperature deformation accumulates more stress concentration and deformation energy storage in the matrix compared with the same process at room or elevated temperature [13]. This energy storage provides a huge impetus for DRX nucleation and atomic diffusion. The phase precipitation and growth are essentially behaviors of the polarization and growth during atomic diffusion [24,25]. Deformation can form a large number of defects, such as dislocations. They can provide more nucleation sites for second-phase precipitation, and can also become fast diffusion channels for the solute atoms and accelerate the atomic diffusion.

According to the results, the sequence of precipitation for the Mg–Zn phase was $\beta_1'(Mg_4Zn_7) \rightarrow \beta_2'(MgZn_2) \rightarrow \beta(MgZn)$ [25,26]. In this experiment, due to the limitation of the detection means, it cannot be determined whether there was a Mg_4Zn_7 or MgZn mixture, and only MgZn$_2$ was detected. Thus, we concluded that MD rolling mainly led to the precipitation of the MgZn$_2$ phase. Under the stress input provided by rolling, solute atoms still have a certain diffusion capacity, so an oversaturated matrix still has the ability to produce a second-phase precipitation. However, the low-temperature environment is not conducive to this reaction. That is, low temperature inhibits the precipitation of an equilibrium precipitation phase [27]. The balance between high energy storage and low temperature conditions causes an incomplete reaction. This prompts the emergence of the non-equilibrium precipitate product. This is why MgZn$_2$ exists as a non-primary phase in the matrix of MD-rolled alloy at cryogenic temperature.

However, it is noteworthy that the presence of the MgZn$_2$ phase was not detected in conventional low-temperature rolling. The type, structure, and shape of the second phase depends on the strain energy inside the alloy [22]. The shear stress in the XY direction inside the alloy during stress loading is shown in Figures 9 and 10. The difference in the shear stress inside the alloy also reflects the difference in the diffusion law of the solute atoms in the alloy [28]. Compared with conventional rolling, there is a stress-loading axis transformation in the MDR process, which increases the crystal defects inside the alloy (dislocations, vacancies, etc.). Such a change raises the energy of the alloy, and the whole system reaches an unstable state. To stabilize the system, a new phase forms cores at the defect positions to release energy, thus reducing the free energy of the system. In other words, MDR makes MgZn$_2$ phase precipitation more prone. The critical radius of nucleation for the precipitated phase is the following [29]:

$$r^* = \frac{2\gamma}{\Delta G_v + \Delta G_\varepsilon + \Delta G_D},$$

where γ is the interface energy between the new phase and the parent phase; ΔG_v is the free energy difference between new phase and the matrix in the unit volume; ΔG_ε is the strain energy of the new phase per unit volume; and ΔG_D is the reduced-system free energy caused by the nucleation of MgZn$_2$ on the defect. Therefore, the increasing of the defect increases ΔG_D, decreases the radius of the critical nucleation, and the number of nuclei increases, so that the driving force of the nuclei increases. This results in the fine-grained size of the precipitated phase. A large number of MgZn$_2$ phases triggers a huge total surface energy. The spherical particles have the lowest surface energy in all shapes, so the precipitated phase gradually tends to globularity after nucleation. Thus, to eliminate the concentrated stress level, a spherical (2H) Laves nano-MgZn$_2$ phase with hcp structure was precipitated in the cryogenic MD-rolled alloy [30]. The coupling of stress direction to crystal orientation reduces the lattice mismatch and solute diffusion rate. In the process of uniform isothermal precipitation, the driving force of the precipitation ΔG can be described by the following formula [31]:

$$\Delta G = -\frac{kT}{V_{at}} \ln\left(\frac{C}{C_{eq}}\right)$$

where V_{at} is the solute atomic volume; C_{eq} is the solubility of MgZn$_2$ in the alloy at equilibrium; C is the current solubility of MgZn$_2$ in the alloy; k is the Bozmann constant; and T is the absolute temperature. Therefore, the lower the T, the smaller the precipitation driving force of the second phase. The second-phase nucleus does not necessarily grow up directly, but only grows up when the size exceeds a certain critical value. At low temperatures, the driving force is insufficient to sustain the MgZn$_2$ phase size. The explosive uniform nuclear barrier and the tendency to coarsening of the precipitated phase are suppressed. Therefore, we ultimately obtained a high-density and uniformly distributed MgZn$_2$ nanoparticle precipitate phase in the cryogenic MDR alloy.

4.2. The Strengthening Effect of MgZn$_2$

MgZn$_2$ hinders the migration of grain boundaries [32,33]. The velocity v of the grain boundary when passing through a single particle obeys the following formula [34]:

$$v = \left(\frac{L}{\sigma_{gb}}\right)\left(\Delta f - \Delta f_{drag}\right)$$

where L is the mobility; σ_{gb} is the grain boundary energy; Δf is driving force; Δf_{drag} is the dragging force. Obviously, in the process of grain boundary migration, the contact area between the precipitated phase and the grain boundary becomes larger and larger. The drag force of particles on grain boundary migration increases, and the speed of grain boundary migration decreases accordingly. When there are multiple second-phase particles, the pinning force F_A received by the grain boundary follows the following formula [34]:

$$F_A = \sigma_{gb} \sin\theta' \left(\frac{2}{d - 2r\cos\theta} - 1\right)$$

The significance of each parameter is indicated in Figure 11. Both the large number of precipitated phases and the fine particle spacing can effectively enhance the hindrance of grain boundary migration. Thus, nano-particles can reduce the peak of the function between grain-size distribution and time. This implies that $MgZn_2$ can effectively impede the disappearance of small-size grains [35]. Therefore, the cryogenic MD-rolled alloy possesses the smallest grains, and the grains still maintain the finest size among the six states after high temperature compression.

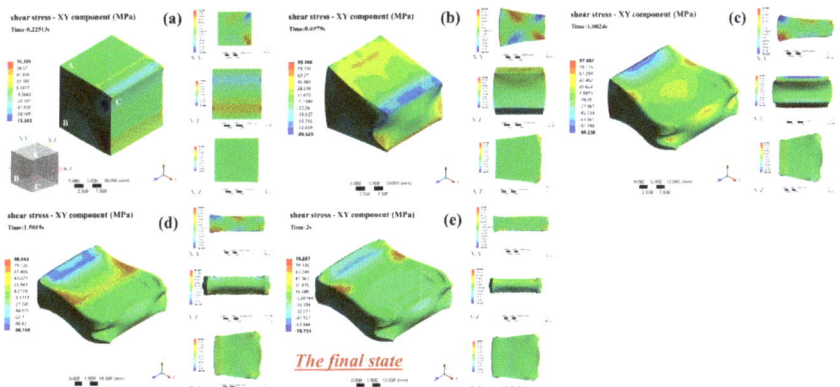

Figure 9. The shear–stress distribution of conventionally rolled alloys. The graphs are cut from different time points of deformation. (**a**) 0.22513 s; (**b**) 0.6979 s; (**c**) 1.0024 s; (**d**) 1.5019 s; (**e**) 2 s.

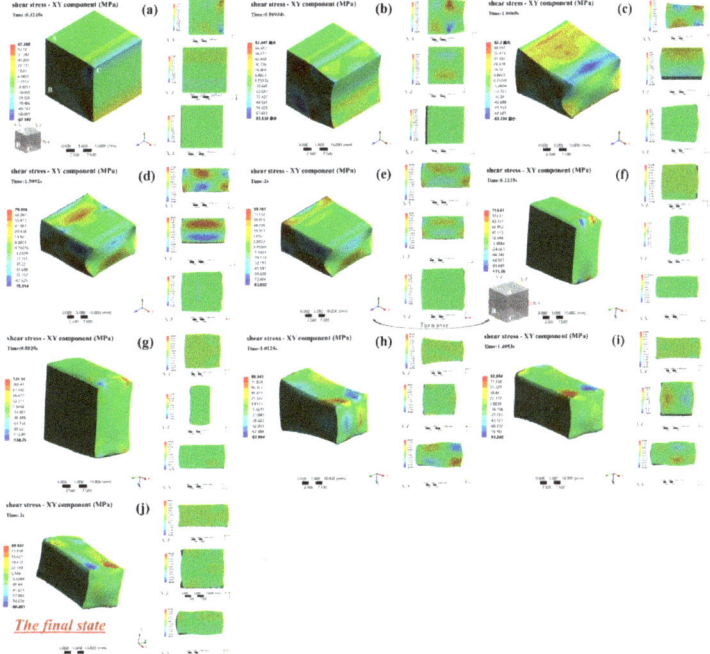

Figure 10. The shear–stress distribution of MD–rolled alloys. The graphs are cut from different time points of deformation. (**f–j**) are redefined according to the time after the specimen is turned over. (**a**) 0.3225 s; (**b**) 0.56934 s; (**c**) 1.0065 s; (**d**) 1.5092 s; (**e**) 2 s; (**f**) 0.2225 s; (**g**) 0.5025 s; (**h**) 1.0125 s; (**i**) 1.4953 s; (**j**) 2 s.

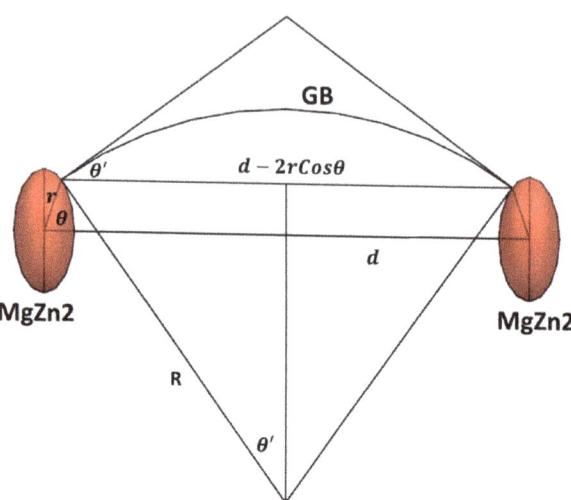

Figure 11. Dragging of the grain boundary migration by two MgZn$_2$ particles.

This nano-precipitated phase is difficult for dislocations to cut and can effectively pin their movement. This promotes the intertwining phenomenon of dislocations [23]. In the cryogenic environment, dislocations cannot be released immediately, which further enhances the number of dislocations and the degree of dislocation packing in the matrix. Thus, dislocation strengthening is further optimized.

5. Conclusions

1. Cryogenic MDR triggers the formation of a large number of nano-grains.
2. The differences in MDR and R deformation lead to shear-stress changes inside the alloy.
3. A large amount of uniformly dispersed nano-precipitation-phase MgZn$_2$ appears only in the cryogenic MDR LZ1641 alloy.
4. MgZn$_2$ has an obstructive effect on the migration of grain boundaries.
5. MgZn$_2$ cannot be cut by dislocations, by which the effect of dislocation strengthening is consolidated.

Author Contributions: Q.J.: Study design, data analysis, data interpretation, writing; X.M.: literature search, data analysis; R.W.: data analysis; S.J.: figures; J.Z.: data collection; L.H.: literature search. All authors have read and agreed to the published version of the manuscript.

Funding: This paper was supported by Ph.D. Student Research and Innovation Fund of the Fundamental Research Funds for the Central Universities (3072020GIP1015), National Natural Science Foundation of China (51871068, 51971071, 52011530025, U21A2049, 52271098), Fundamental Research Funds for the Central Universities (3072022QBZ1002), National Key Research and Development Program of China (2021YFE0103200).

Data Availability Statement: Not applicable.

Conflicts of Interest: The authors declare no conflict of interest.

References

1. Wang, J.H.; Xu, L.; Wu, R.Z.; Feng, J.; Zhang, J.H.; Hou, L.G.; Zhang, M.L. Enhanced Electromagnetic Interference Shielding in a Duplex-Phase Mg-9Li-3Al-1Zn Alloy Processed by Accumulative Roll Bonding. *Acta Metall. Sin. (Engl. Lett.)* **2020**, *33*, 490–499. [CrossRef]
2. Li, C.Q.; Xu, D.K.; Zu, T.T.; Han, E.H.; Wang, L. Effect of temperature on the mechanical abnormity of the quasicrystal reinforced Mg–4%Li–6%Zn–1.2%Y alloy. *J. Magnes. Alloys* **2015**, *3*, 106–111. [CrossRef]

3. Huang, Z.H.; Wang, L.Y.; Zhou, B.J.; Fischer, T.; Yi, S.B.; Zeng, X.Q. Observation of non-basal slip in Mg-Y by in situ three-dimensional X-ray diffraction. *Scr. Mater* **2018**, *143*, 44–48. [CrossRef]
4. Li, B.; Hou, L.; Wu, R.; Zhang, J.; Li, X.; Zhang, M.; Dong, A.; Sun, B. Microstructure and thermal conductivity of Mg-2Zn-Zr alloy. *J. Alloys Compd.* **2017**, *722*, 772–777. [CrossRef]
5. Fei, P.; Qu, Z.; Wu, R. Microstructure and hardness of Mg–9Li–6Al–xLa (x=0, 2, 5) alloys during solid solution treatment. *Mater. Sci. Eng. A* **2015**, *625*, 169–176. [CrossRef]
6. Mahmudi, R.; Shalbafi, M.; Karami, M.; Geranmayeh, A.R. Effect of Li content on the indentation creep characteristics of cast Mg-Li-Zn alloys. *Mater. Des.* **2015**, *75*, 184–190. [CrossRef]
7. Karewar, S.; Gupta, N.; Groh, S.; Martinez, E.; Caro, A.; Srinivasan, S.G. Effect of Li on the deformation mechanisms of nanocrystalline hexagonal close packed magnesium. *Comp. Mater. Sci.* **2017**, *126*, 252–264. [CrossRef]
8. Zhang, H.; Hao, H.L.; Fu, G.Y.; Liu, B.S.; Li, R.G.; Wu, R.Z.; Pan, H.C. Microstructure and Mechanical Property of Hot-Rolled Mg-2Ag Alloy Prepared with Multi-pass Rolling. *Acta Metall. Sin. (Engl. Lett.)* **2022**. [CrossRef]
9. Alam, M.; Groh, S. Dislocation modeling in bcc lithium: A comparison between continuum and atomistic predictions in the modified embedded atoms method. *J. Phys. Chem. Solids* **2015**, *82*, 1–9. [CrossRef]
10. Zeng, Z.R.; Stanford, N.; Davies, C.H.J.; Nie, J.F.; Birbilis, N. Magnesium extrusion alloys: A review of developments and prospects. *Int. Mater. Rev.* **2019**, *64*, 27–62. [CrossRef]
11. Liu, X.; Du, G.; Wu, R.; Niu, Z.; Zhang, M. Deformation and microstructure evolution of a high strain rate superplastic Mg-Li-Zn alloy. *J. Alloys Compd.* **2011**, *509*, 9558–9561. [CrossRef]
12. Liu, X.-Y.; Capolungo, L.; Hunter, A. Screw dislocation impingement and slip transfer at fcc-bcc semicoherent interfaces. *Scr. Mater* **2021**, *201*, 113977. [CrossRef]
13. Ji, Q.; Zhang, S.; Wu, R.; Jin, S.; Zhang, J.; Hou, L. High strength BCC magnesium-Lithium alloy processed by cryogenic rolling and room temperature rolling and its strengthening mechanisms. *Mater. Charact.* **2022**, *187*, 111869. [CrossRef]
14. Wang, B.J.; Xu, D.K.; Cai, X.; Qiao, Y.X.; Sheng, L.Y. Effect of rolling ratios on the microstructural evolution and corrosion performance of an as-rolled Mg-8 wt.%Li alloy. *J. Magnes. Alloys* **2021**, *9*, 560–568. [CrossRef]
15. Hu, L.F.; Gu, Q.F.; Li, Q.; Zhang, J.Y.; Wu, G.X. Effect of extrusion temperature on microstructure, thermal conductivity and mechanical properties of a Mg-Ce-Zn-Zr alloy. *J. Alloys Compd.* **2018**, *741*, 1222–1228. [CrossRef]
16. Jin, S.; Liu, H.; Wu, R.; Zhong, F.; Hou, L.; Zhang, J. Combination effects of Yb addition and cryogenic-rolling on microstructure and mechanical properties of LA141 alloy. *Mater. Sci. Eng. A* **2020**, *788*, 139611. [CrossRef]
17. Li, C.; Deng, B.; Dong, L.; Liu, X.; Du, K.; Shi, B.; Dong, Y.; Peng, F.; Zhang, Z. Effect of Zn addition on the microstructure and mechanical properties of as-cast BCC Mg-11Li based alloys. *J. Alloys Compd.* **2022**, *895*, 162718. [CrossRef]
18. Ji, Q.; Wang, Y.; Wu, R.; Wei, Z.; Ma, X.; Zhang, J.; Hou, L.; Zhang, M. High specific strength Mg-Li-Zn-Er alloy processed by multi deformation processes. *Mater. Charact.* **2020**, *160*, 110135. [CrossRef]
19. Yuan, T.; Wu, Y.; Liang, Y.; Jiao, Q.; Zhang, Q.; Jiang, J. Microstructural control and mechanical properties of a high Li-containing Al-Mg-Li alloy. *Mater. Charact.* **2021**, *172*, 110895. [CrossRef]
20. Qin, S.; Lee, S.; Tsuchiya, T.; Matsuda, K.; Horita, Z.; Kocisko, R.; Kvackaj, T. Aging behavior of Al-Li-(Cu, Mg) alloys processed by different deformation methods. *Mater. Des.* **2020**, *196*, 109139. [CrossRef]
21. Shao, H.; Huang, Y.; Liu, Y.; Xiao, Z. Structural stability, anisotropic elasticities and electronic structure of η-MgZn$_2$ under pressures: A first-principle investigation. *Solid State Commun.* **2022**, *343*, 114644. [CrossRef]
22. Liu, H.; Nie, J.F. Phase field simulation of microstructures of Mg and Al alloys. *Mater. Sci. Technol.* **2017**, *33*, 2159–2172. [CrossRef]
23. Zou, Y.; Zhang, L.H.; Li, Y.; Wang, H.T.; Liu, J.B.; Liaw, P.K.; Bei, H.B.; Zhang, Z.W. Improvement of mechanical behaviors of a superlight Mg-Li base alloy by duplex phases and fine precipitates. *J. Alloys Compd.* **2018**, *735*, 2625–2633. [CrossRef]
24. Zhou, B.-C.; Shang, S.-L.; Wang, Y.; Liu, Z.-K. Diffusion coefficients of alloying elements in dilute Mg alloys: A comprehensive first-principles study. *Acta Mater.* **2016**, *103*, 573–586. [CrossRef]
25. Wang, D.; Amsler, M.; Hegde, V.I.; Saal, J.E.; Issa, A.; Zhou, B.-C.; Zeng, X.; Wolverton, C. Crystal structure, energetics, and phase stability of strengthening precipitates in Mg alloys: A first-principles study. *Acta Mater.* **2018**, *158*, 65–78. [CrossRef]
26. Zhong, Q.; Pan, D.; Zuo, S.; Li, X.; Luo, H.; Lin, Y. Fabrication of MgZn intermetallic layer with high hardness and corrosion resistance on AZ31 alloy. *Mater. Charact.* **2021**, *179*, 111365. [CrossRef]
27. Liu, Y.; Shao, S.; Xu, C.; Zeng, X.; Yang, X. Effect of cryogenic treatment on the microstructure and mechanical properties of Mg-1.5Zn-0.15Gd magnesium alloy. *Mater. Sci. Eng. A* **2013**, *588*, 76–81. [CrossRef]
28. Nagarajan, D.; Caceres, C.H. The friction stress of the Hall-Petch relationship of pure Mg and solid solutions of Al, Zn, and Gd. *Metall. Mater. Trans. A* **2018**, *49*, 5288–5297. [CrossRef]
29. Doherty, R.D. Chapter 15—Diffusive phase transformations in the solid state. In *Physical Metallurgy*, 4th ed.; Cahn, R.W., Haasen, P., Eds.; North-Holland: Oxford, UK, 1996; pp. 1363–1505.
30. Zheng, B.; Zhao, L.; Hu, X.B.; Dong, S.J.; Li, H. First-principles studies of Mg$_{17}$Al$_{12}$, Mg$_2$Al$_{13}$, Mg$_2$Sn, MgZn$_2$, Mg$_2$Ni and Al$_3$Ni phases. *Phys. B* **2019**, *560*, 255–260. [CrossRef]
31. Aaronson, H.I.; Kinsman, K.R.; Russell, K.C. The volume free energy change associated with precipitate nucleation. *Scr. Metall.* **1970**, *4*, 101–106. [CrossRef]
32. Ahmed, K.; Tonks, M.; Zhang, Y.; Biner, B.; El-Azab, A. Particle-grain boundary interactions: A phase field study. *Comp. Mater. Sci.* **2017**, *134*, 25–37. [CrossRef]

33. Peng, H.R.; Liu, W.; Hou, H.Y.; Liu, F. Pinning effect of coherent particles on moving planar grain boundary: Theoretical models and molecular dynamics simulations. *Materialia* **2019**, *5*, 100225. [CrossRef]
34. Chang, K.; Feng, W.; Chen, L.-Q. Effect of second-phase particle morphology on grain growth kinetics. *Acta Mater.* **2009**, *57*, 5229–5236. [CrossRef]
35. Huber, L.; Rottler, J.; Militzer, M. Atomistic simulations of the interaction of alloying elements with grain boundaries in Mg. *Acta Mater.* **2014**, *80*, 194–204. [CrossRef]

Article

Evolution of the Microstructure and Mechanical Properties of AZ31 Magnesium Alloy Sheets during Multi-Pass Lowered-Temperature Rolling

Qing Miao [1], Lantao Zhu [2], Wenke Wang [2,*], Zhihao Wang [3], Bin Shao [4], Wenzhen Chen [2], Yang Yu [2] and Wencong Zhang [2]

1. School of Materials, Shanghai Dian Ji University, Shanghai 201306, China
2. School of Materials Science and Engineering, Harbin Institute of Technology, Weihai 264209, China
3. Laboratoire de Génie Civil et Génie Mécanique—EA 3913, INSA Rennes, University Rennes, F-35000 Rennes, France
4. National Key Laboratory for Precision Hot Pressing of Metals, Harbin Institute of Technology, Harbin 150001, China
* Correspondence: 15b309020@hit.edu.cn; Tel.: +86-631-567-2167; Fax: +86-631-567-2167

Abstract: AZ31 magnesium alloy sheets with 2 mm thickness were successfully fabricated by multi-pass lowered-temperature rolling. The evolution of the microstructure, texture, and mechanical properties during the rolling process was investigated. Based on the effect of multiple dynamic recrystallization, multi-pass lowered-temperature rolling not only refined the grain size obviously but also markedly improved the microstructure homogeneity. The resulting sheets had the optimal microstructure morphology with an average grain size of 4.38 µm. For the texture evolution, the stress state of the rolling process made the (0002) basal plane gradually rotate toward the rolling plane. However, the activation of non-basal slips due to the higher rolling temperature slightly rotated the (0002) basal plane point to the rolling direction (RD). As a result, the grain refinement strengthening and the texture strengthening together increased the yield stress to 202 MPa in the transverse direction (TD) and 189.8 MPa in the RD. Importantly, the resulting sheet concurrently exhibited excellent fracture elongation, about 38% in the TD and 39.2% in the RD. This was mainly ascribed to the finer grain size, giving rise to a significant effect of grain boundary sliding and the activation amount of non-basal slips.

Keywords: magnesium alloy; dynamic recrystallization; grain size; texture; elongation

1. Introduction

Magnesium alloys are considered to be the lightest structural metallic alloys, resulting in marked increases in their use in automobile parts and electric appliance cases [1–3]. In magnesium alloy products, magnesium alloy sheets have a very large market demand. However, the hexagonal close-packed (HCP) structure in magnesium alloys only activates two independent basal slips at room temperature [4]. This results in their poor formability, which is an important factor restricting their wide use in critical safety components [5]. In order to meet the demand, it is necessary to develop rolling technologies for the mass production of magnesium alloy sheets with high performance (high strength and high formability simultaneously).

It is well known that grain refinement and texture weakening are factors that improve the formability of magnesium alloy sheets [6–8]. Huang et al. fabricated various AZ31 magnesium alloy sheets with a grain size from 7 µm to 16 µm and found that both stretch formability and deep drawability deteriorated with the increase in grain size [7]. They attributed the reason to the small grain size that restricted tension twin activity and finally delayed the texture strengthening [7]. In Huo's study, the tensile ductility and

stretch formability of AZ31 magnesium alloy sheets fabricated by cross-wavy bending were distinctly enhanced compared to the initial sheets (about 1.55 and 2 times, respectively) [9]. In their view, these prominent increases were mainly ascribed to the fine-grained microstructure with an average grain size of 8 μm and a weak and random basal texture [9]. Currently, abundant plastic process techniques can refine the grain size and tailor the basal texture synchronously, for example, multi-pass lowered-temperature rolling [10], twin-roll casting [11], different speed rolling [12], and equal channel angular rolling [13]. However, these techniques cannot realize grain refinement and texture weakening or the high efficiency of industrial sheet fabrication. In the above process methods, multi-pass lowered-temperature rolling is considered to be the most suitable method for industrial sheet fabrication due to the absence of intermediate annealing [10]. This is very important to the mass production of magnesium alloy sheets. Moreover, prior work verified that multi-pass lowered-temperature rolling could fabricate magnesium alloy sheets with high performance [14,15]. Such an effect might be ascribed to the occurrence of multiple dynamic recrystallization (DRX), which helped to refine the grain size and improve the microstructure homogeneity during the rolling process. Accordingly [16], the DRX grain size was closely related to the rolling process parameters, which can be well explained by the Zener–Hollomon Z ($Z = \dot{\varepsilon} \exp(Q/RT)$, where $\dot{\varepsilon}$ is the strain rate, T is the deformation temperature, Q is the activation energy, and R is the gas constant). Actually, the strategy of a lower temperature was designed during multi-pass rolling, which was because such a strategy can increase the recrystallization nucleation sites and inhibit the grain boundary migration. Therefore, multi-pass lowered-temperature rolling can significantly improve the microstructure of magnesium alloys. However, the present rolling technologies usually use thick magnesium alloy plates as the initial material, which need multiple rolling passes to realize the fabrication of thin sheets, i.e., only by reasonable matching of the rolling parameters under each rolling pass can the microstructure of the resulting sheets be improved as much as possible. This makes it necessary to investigate the evolution of the microstructure and mechanical properties during multi-pass hot rolling. For this purpose, this work investigated the multi-pass lowered-temperature rolling of as-cast AZ31 magnesium alloy sheets in order to fabricate thin sheets with better microstructure and high performance. The reason for the enhancement of elongation arising from grain refinement was mainly discussed.

2. Materials and Methods

The initial material in this work was the as-cast AZ31 magnesium alloy plates with 30 mm thickness. Their nominal composition was determined as Mg, 3 wt.%; Al, 1 wt.%; Zn, 0.2 wt.%; and Mn by an inductively coupled plasma analyzer (ICP, Thermo Corp., Waltham, MA, USA). These initial materials were homogenized at 400 °C for 24 h in an electric furnace to improve component segregation and eliminate possible intercrystalline phases according to the literature [17]. Moreover, in order to improve their workability, all sheets were preheated to 400 °C for 30 min prior to hot rolling. In this work, all thickness reductions were set to 20% at break-down rolling pass in order to avoid premature cracking and then increased to 30% to improve the degree of DRX. According to previous studies, a lower deformation temperature easily realizes grain refinement. Therefore, in order to refine the grain size effectively, the rolling temperature tended to decrease as the rolling proceeded, i.e., 400 °C (Pass 1), 380 °C (Pass 2), 360 °C (Pass 3), 340 °C (Pass 4), 320 °C (Pass 5), 300 °C (Pass 6), 280 °C (Pass 7), and 260 °C (Pass 8). The detailed rolling process parameters are summarized in Table 1.

Table 1. Rolling process parameters for AZ31 magnesium alloy.

Condition	Thickness Change, mm	Rolling Temperature, °C	Thickness Reduction, %
Pass 1	30 → 24	400	20
Pass 2	24 → 16.8	380	30
Pass 3	16.8 → 11.7	360	30
Pass 4	11.7 → 8.2	340	30
Pass 5	8.2 → 5.7	320	30
Pass 6	5.7 → 4	300	30
Pass 7	4 → 2.8	280	30
Pass 8	2.8 → 2	260	30

Microstructure characteristics were observed on the RD–TD plane (RD: rolling direction, TD: transverse direction) of the samples, which were machined from the sheets by wire-electrode cutting. These samples first experienced mechanical polishing and then chemical etching with a solution consisting of picric acid (5.5 g), acetic acid (2 mL), distilled water (10 mL), and ethanol (90 mL). Then, an OLYMPUS GX71 (Olympus Corp., Tokyo, Japan) optical microscope (OM) was applied for metallographic observations. In this work, the mean linear intercept method ($\overline{d_v} = 1.74\ \overline{L}$, $\overline{L}$ is the average linear intercept) was used to measure the average grain size. Regarding the texture measurement, the texture state of the initial material was characterized by X-ray diffraction (XRD) experiments and that of the as-rolled sheets by electron backscatter diffraction (EBSD). The preparation of EBSD samples consisted of soft diamond polishing and subsequent etching with electropolishing in a 5:3 solution of ethanol and phosphoric acid for 8 min at 0.25 A. Subsequently, these samples were measured by a JEOL 733 electron probe (JEOL Ltd., Tokyo, Japan) equipped with an HKL Channel 5 system. In this work, scanning electron microscopy (SEM) was used to identify the type of fracture of the tension samples. In addition, a TECNAL transmission electron microscopy (TEM, FEI Corp., Hillsboro, OR, USA) was used to characterize the microstructure details, and its accelerating voltage was 200 kV. Uniaxial tension tests were measured by an Instron 5569 machine (Instron Corp., Norwood, MA, USA) with a strain rate of $1 \times 10^{-3}\ s^{-1}$ at room temperature. The tension samples were cut along the rolling direction and the transverse direction and their gauges were 25 mm in length and 6 mm in width.

3. Results and discussion
3.1. Microstructure Characteristics

Figure 1 shows the microstructure of the initial material after homogenized annealing. Clearly, the initial material possessed a larger grain size of about 293.3 μm, but relatively speaking, the microstructure was homogeneous, demonstrating that the difference in grain size between RD and TD was small. The microstructure characteristics of as-rolled AZ31 magnesium alloy sheets during multi-pass lowered-temperature rolling are displayed in Figure 2, and their corresponding grain size distributions are summarized in Figure 3. Clearly, the microstructure in the Pass 1 sheet exhibited a combination of fine and coarse grains, but compared to the initial microstructure, significant grain refinement occurred, in which the average grain size decreased to 34.73 μm in the Pass 1 sheet (Figure 3a). In addition, abundant twins could be seen in coarse grains (Figure 2a). In the Pass 2 sheet, this microstructure feature (coarse grains containing twins) was still the dominant, but the coarse grains were surrounded by fine, equiaxed grains (Figure 2b). This revealed the DRX had occurred. Here, it is worth noting the reason for the activation of twinning in the initial stage of rolling. The coarse grain in the initial material was considered to be the major cause. More specifically, in coarse grains, the distance of the dislocation slip was longer, which generally led to severe dislocation pile-up and stress concentration at grain boundaries [18]. Therefore, such a stress concentration would promote the twinning activation. In addition, the crystal orientation was an important factor as well. The initial material in this work was the as-cast AZ31 magnesium alloy plates, and there were some grains with the c-axes nearly perpendicular to the normal direction (ND) of the plate. Such crystal orientation was conducive to the activation of tension twinning when the compression force imposed on the plates occurred along the normal direction. However, as the rolling proceeded, abundant basal planes were rotated to the hard orientation. At this moment, the tension twinning lost its original advantage of crystal orientation and thereby was activated with difficulty. As a consequence, the coarse grain and the crystal orientation were two important reasons triggering the twinning activation, particularly at the initial stage of rolling. Importantly, for the magnesium alloys, the activation of twinning in the initial stage of rolling had two beneficial roles: (1) they could accommodate the rolling deformation and release stress concentration, avoiding premature cracking; (2) their twin boundaries could act as the preferred sites for DRX in the next rolling procedure, and this could be a softening mechanism to improve the workability, ensuring that subsequent rolling was done smoothly [18].

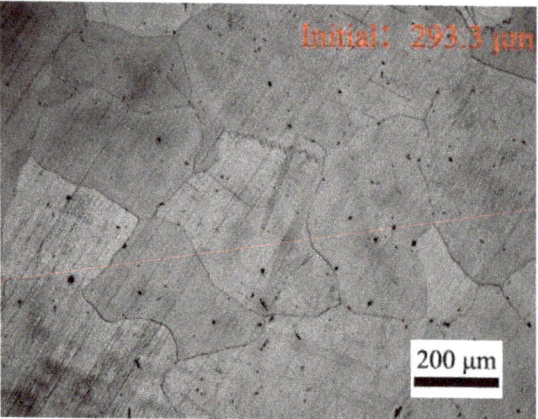

Figure 1. Metallographic observation of the initial material after homogenized annealing.

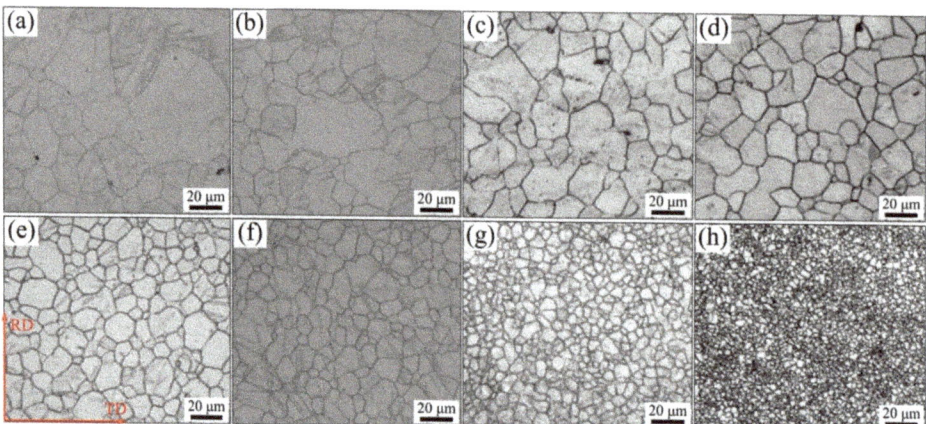

Figure 2. Microstructure characteristics of AZ31 magnesium alloy sheets during multi-pass lowered-temperature rolling: (**a**) Pass 1; (**b**) Pass 2; (**c**) Pass 3; (**d**) Pass 4; (**e**) Pass 5; (**f**) Pass 6; (**g**) Pass 7; (**h**) Pass 8. (RD: rolling direction, TD: transverse direction).

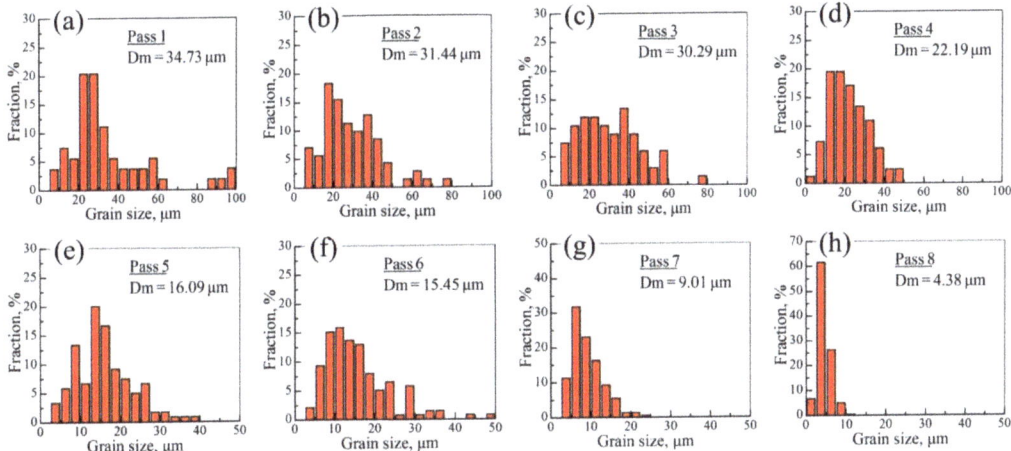

Figure 3. Grain size distributions of AZ31 magnesium alloy sheets during multi-pass lowered-temperature rolling: (**a**) Pass 1; (**b**) Pass 2; (**c**) Pass 3; (**d**) Pass 4; (**e**) Pass 5; (**f**) Pass 6; (**g**) Pass 7; (**h**) Pass 8 (Dm refers to the average grain size).

Further decreasing the rolling deformation increased the degree of DRX, producing a large number of fine, equiaxed grains in the Pass 3 sheet and Pass 4 sheet (Figure 2). Figure 3 shows that their average grain sizes were refined to 30.29 μm in the Pass 3 sheet and 22.19 μm in the Pass 4 sheet. However, from their grain size distributions, 7.27–75.7 μm in the Pass 3 sheet and 3.53–46.21 μm in the Pass 4 sheet, it could be concluded that their microstructure homogeneity was not good enough [19]. Moreover, their coarse grains were highly distorted, suggesting that the new DRX embryo was developing at the grain boundaries [20]. This further demonstrated their mixed microstructure with large and fine grains. After Passes 5 and 6, the abnormally coarse grain had nearly disappeared and the fractions of DRX grain obviously increased (Figure 2e,f). The microstructure homogeneity was distinctly improved compared to the prior microstructure homogeneity, which could be supported by the unimodal grain size distributions (Figure 3e,f). After Pass 7, the microstructure presented a typical DRX microstructure appearance with excellent microstructure homogeneity. The resulting sheet after Pass 8 had the smallest average grain size of 4.38 μm (Figure 3h). The above metallographic results suggested that, based on the principle of multiple DRX, multi-pass lowered-temperature rolling was not only beneficial to refine the grain size but also to improve the microstructure homogeneity. In addition, in order to obtain a better magnesium alloy microstructure in the future, mathematical modeling between the rolling parameters and the DRX grain size should be established based on the Zener–Hollomon law, similar to the mathematical modeling established in the literature [21].

In order to further characterize the DRX behavior during the rolling, the TEM observation and the grain orientation spread (GOS) map was applied for the Pass 8 sheet, as shown in Figure 4. Clearly, Figure 4a shows that each grain interior was clean and uniform. This suggested that no crystal orientation difference occurred in one individual grain and thereby no obvious defects developed in such grains. Additionally, Figure 4b shows the DRX microstructure of the Pass 8 sheet using the GOS map in which the grains were outlined by the high-angle grain boundaries ranging from 15° to 180°, and the grains with the GOS values smaller than 2° were considered to be the DRX grains (blue and green color) according to the literature [22,23]. Obviously, there were only blue- and green-colored grains in the microstructure, again validating the complete DRX behavior in the Pass 8 sheet. These results together demonstrated that the dislocation density was lower due to the complete DRX behavior during the multi-pass lowered-temperature rolling.

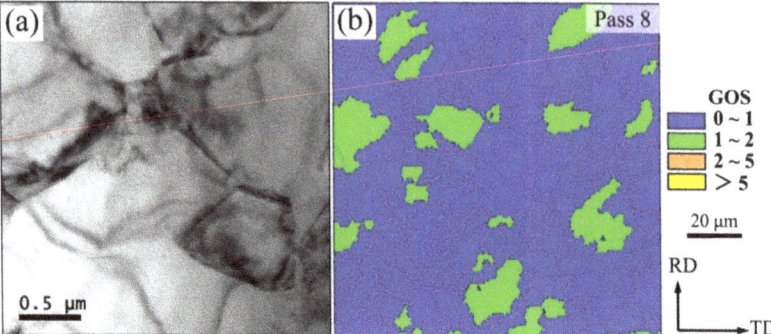

Figure 4. The TEM observations (**a**) and the grain orientation spread (GOS) map for the Pass 8 sheet (**b**). (In GOS map, the GOS value in each grain was determined by calculating the average misorientation between all points within the grain).

3.2. Texture Characteristics

It is well known that the HCP structure in magnesium alloy is liable to cause a strong (0002) basal texture, strongly influencing the mechanical properties [24,25]. Therefore, it was necessary to investigate the texture evolution in multi-pass lowered-temperature rolling. According to the literature [16], the texture state of magnesium alloy can be estimated by both XRD and EBSD. In this work, XRD was applied to measure the texture state of the initial material, and the ratios of the diffraction peaks, $I_{10\bar{1}1}/I_{10\bar{1}0}$ and $I_{10\bar{1}1}/I_{0002}$, were calculated to quantify the texture characteristics. Figure 5a,b shows the XRD spectrum of the initial material and pure magnesium powder (PDF#35-0821), respectively. Clearly, the ratios of the initial material were similar to those of the standard magnesium card. This suggested that there was no preferred crystalline orientation in the initial material.

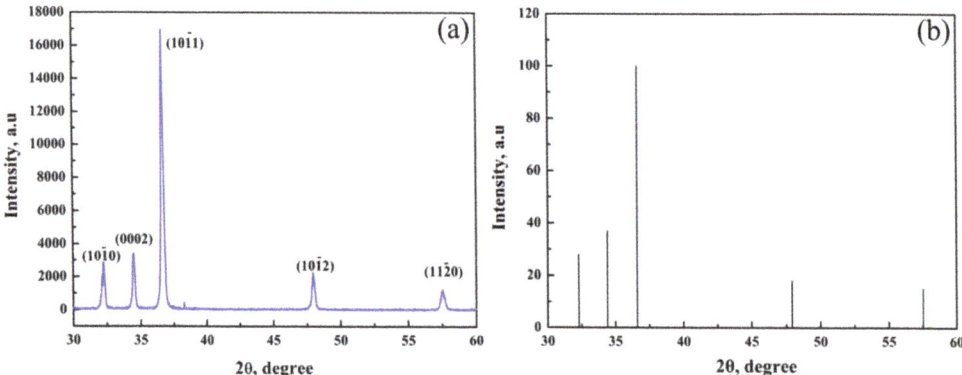

Figure 5. X-ray diffraction spectra of the initial material: (**a**) cast, (**b**) standard spectra.

With regard to the as-rolled sheets, their texture states were measured by EBSD. Figure 6 shows the (0002), (10–10), and (11–20) pole figures under different rolling passes. After one pass, there were multiple peaks in the (0002) pole figures (Figure 6a), and the maximum texture intensity reached 6.01. This demonstrated that the texture morphology emerged in the Pass 1 sheet. After two passes, the peak was not only strengthened but also rotated to the center of the (0002) pole figure (Figure 6b), i.e., more and more c-axes were parallel to the normal direction (ND), and this made the maximum texture intensity increase to 6.5. Wang and Huang pointed out that the stress state under rolling conditions made the slip plane rotate toward the rolling plane and the slip direction toward the rolling direction [26]. In magnesium alloys, basal slip is the main deformation mode due to its lower critical resolved shear stress (CRSS) [27]. Therefore, the basal plane would gradually rotate to the RD–TD plane under the combined effect of tension along the RD and compression along the ND. However, it can be seen from Figure 6c–e that there were some deflections from the peak to the ND. This could be explained well by the activation of various slip systems. As reported in [28], the CRSS of non-basal slip will decrease sharply when the deformation temperature is higher than 225 °C. In this work, the rolling temperature was always in excess of 260 °C, i.e., non-basal slip (prismatic slip or pyramidal slip) would be activated and take part in the texture formation. According to the above literature [26], these non-basal planes would gradually rotate to the RD–TD plane, and this would rotate the (0002) basal plane away from the RD–TD plane, but due to the tension effect in the RD, the (0002) basal plane only pointed to the RD. Nevertheless, the basal slip was still the dominant deformation mode when the rolling temperature was higher than 260 °C. Such texture evolution could also be characterized by the inverse pole figure (IPF) map, and here the IPF maps for the Pass 4 and Pass 8 sheets were applied, as shown in Figure 7. In these IPF maps, the color code within the unit triangle represented the orientation of the sample normal direction within the hexagonal unit triangle. Clearly, mainly yellow- and red-colored grains were seen in the Pass 4 sheet (Figure 7a). This was consistent with the characteristics in the (0002) pole figure of the Pass 4 sheet (Figure 6c). By comparison, the red-colored grains significantly increased in the Pass 8 sheet, meaning that the texture component whose c-axes were parallel to the ND was strengthened. Therefore, as the rolling proceeded, the overall trend in the (0002) pole figures was still that more and more c-axes pointed to the ND (Figure 6), and the maximum texture intensity gradually increased.

3.3. Mechanical Properties

Figure 8 shows the measured stress–strain curves of the AZ31 magnesium alloy sheets along the RD and the TD in the tension test under different conditions. Table 2 summarizes the corresponding data for the yield stress (YS, MPa), ultimate tensile stress (UTS, MPa), and fracture elongation (FE, %). In order to clearly characterize the evolution of mechanical properties, the variations in YS, UTS, and FE as a function of the rolling pass are displayed in Figure 9. Clearly, the YS and UTS gradually increased with the increase in the rolling pass. As commonly reported [29], the YS of magnesium alloy sheets had a close relationship with the grain size and the texture state. The influence of grain size on the YS followed the Hall–Petch relationship, while the influence of the texture state was based on the Schmid law [29]. Accordingly, the finer grain size and the intense texture state usually led to a higher YS. As previously described, the average grain size retained a refinement trend, while the (0002) basal texture had a strengthening trend. Therefore, in this work, with the increase in the rolling pass, the increasing trend of YS was ascribed to the grain refinement strengthening and the texture strengthening, as shown in Figure 9. In addition, it should be noted that the planar anisotropy in stress (YS and UTS) between the RD and the TD was small (Figure 9 and Table 2). This reason could be ascribed to two points: first, the new DRX grains generally exhibited equiaxed morphology, indicating the similar grain size along any random direction in the sheet; second, owing to the non-texture state in the initial material, the (0002) basal plane would rotate toward the RD–TD plane along any random direction in the sheet, and this led to the small difference in the texture state between the RD and the TD. Therefore, the similar microstructure state was the main reason for the small planar anisotropy between the RD and the TD. In addition, the sheets during the multi-pass lowered-temperature rolling underwent DRX completely, exhibited a lower dislocation density. This made the contribution from the dislocation density to the strength improvement small, which was not considered in this work. Similarly, in AZ31 magnesium alloy, the contribution of solid solution strengthening to the strength was much smaller compared with the fine-grained strengthening, which was estimated in the literature [30]. As a consequence, this work did not discuss the solid solution strengthening.

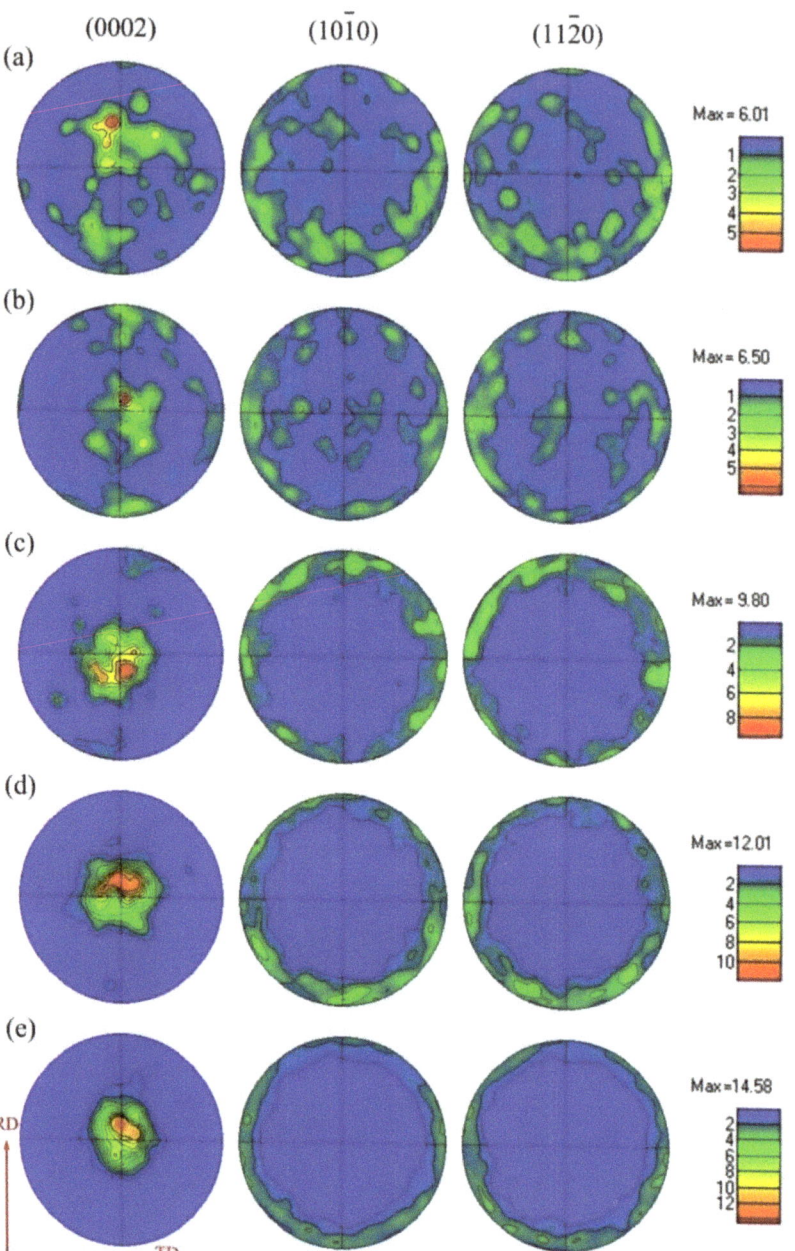

Figure 6. (0002), (10–10), and (11–20) pole figures of AZ31 magnesium alloy sheets during multi-pass lowered-temperature rolling: (**a**) Pass 1; (**b**) Pass 2; (**c**) Pass 4; (**d**) Pass 6; (**e**) Pass 8. (RD: rolling direction, TD: transverse direction).

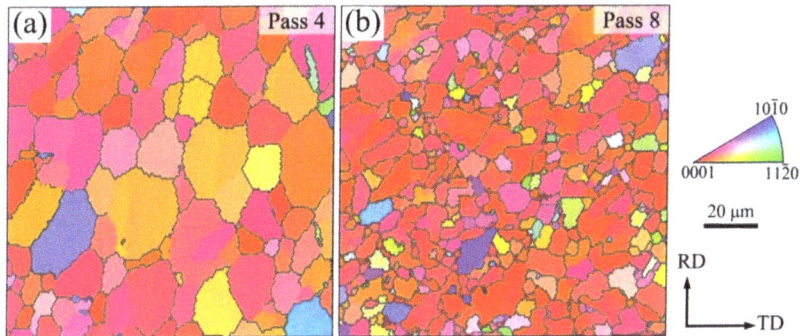

Figure 7. Microstructure characteristics using inverse pole figure maps for the (**a**) Pass 4 sheet and (**b**) Pass 8 sheet.

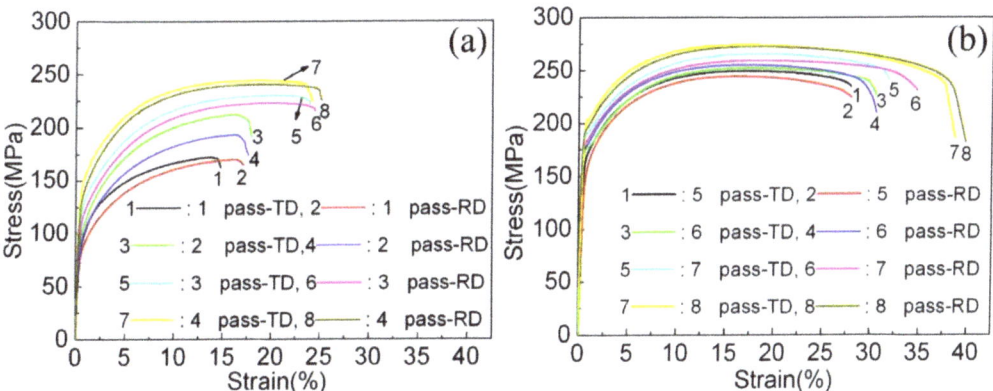

Figure 8. Room temperature stress–strain curves of AZ31 magnesium alloy sheets along the RD and TD during multi-pass lowered-temperature rolling: (**a**) from 1 pass to 4 pass and (**b**) from 5 pass to 8 pass.

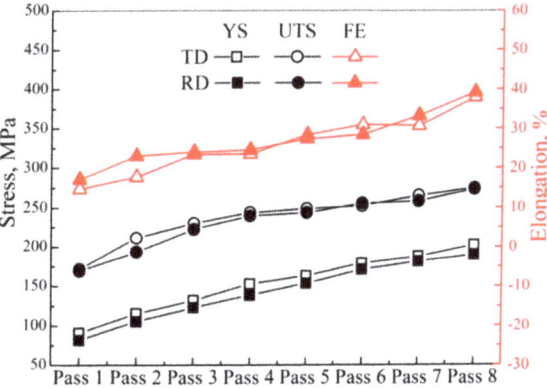

Figure 9. Stress (YS and UTS) and elongation (FE) variation of AZ31 magnesium alloy sheets along the RD and TD during multi-pass lowered-temperature rolling (YS: yield stress; UTS: ultimate tension stress; FE: fracture elongation).

Table 2. Summarized mechanical properties of AZ31 magnesium alloy sheets (YS: yield stress; UTS: ultimate tension stress; FE: fracture elongation).

Condition	YS, MPa		UTS, MPa		FE, %	
	TD	RD	TD	RD	TD	RD
Initial	40		160		7	
Pass 1	90.7	81.6	171.6	169.2	14.7	17.1
Pass 2	115.3	105.2	211.4	193.5	17.7	23
Pass 3	131.9	122.8	230.2	222.6	23.4	24
Pass 4	152.7	138.5	243.5	239.4	23.5	24.5
Pass 5	163.2	153.6	248.9	243.7	28.4	27.4
Pass 6	179.1	171.4	251.8	255.1	31	28.4
Pass 7	188	181.6	264.8	257.9	30.7	33.3
Pass 8	202	189.8	274.3	272.6	38	39.2

For magnesium alloy sheets, poor ductility and formability are the main factors limiting their widespread commercial usability. Therefore, this work focused on the evolution of fracture elongation. Clearly, in both Figure 9 and Table 2, the fracture elongation gradually increased with the increase of the number of rolling passes. The sheets rolled after eight passes exhibited higher fracture elongation (38% along the TD and 39.2% along the RD). To the authors' knowledge, this is not common in previous literature. The main reason was considered to be the finer grain size, which would be carefully discussed from the aspects of metallographic observations, TEM characteristics, and fracture SEM characteristics.

It is well known that a finer grain size has a positive effect on elongation by affecting the contribution of grain boundary sliding [31]. This is supported by Koike, who pointed out that the ability of grain boundary sliding increased with the decline of the grain size, and the contribution to fracture elongation reached 8% when the grain size was less than 8 μm [31]. Many studies also demonstrated that grain boundary sliding was one important factor improving fracture elongation for AZ31 and AZ61 magnesium alloys with grain sizes less than 5 μm [32,33]. Zheng et al. pointed out that the unique combinations of strength and ductility could be realized in bulk polycrystalline pure magnesium by tuning the predominant deformation mode and suggested that the grain boundary sliding governed the plastic deformation in the ultra-fine grain specimen, leading to softening of the material and exceptionally large room temperature tensile elongation up to 65% [34]. In this work, the average grain size of the resulting sheet (Pass 8 sheet) was approximately 4.38 μm, which was conducive to the occurrence of grain boundary sliding. Therefore, grain boundary sliding arising from a finer grain size contributed to the higher fracture elongation of the Pass 8 sheet. In addition, the homogeneous microstructure should be another influential factor. This could be well supported by the metallographic observations of the uniform deformation part and necking part (the part is indicated by the black point in Figure 11a). Clearly, in the uniform deformation part (Figure 10a), most of the grains had equiaxed morphology. For the necking part (Figure 10b), although the grains were elongated along tension deformation, the microstructure was still relatively homogeneous and there was no crack nucleation point. Here, their homogeneous microstructure was beneficial to accommodate the tension deformation between grains and to release the stress concentration around the grain boundaries. This prevented the crack appearing prematurely, which further improved their fracture elongation.

Another mechanism describing the higher fracture elongation of the resulting sheets is the activation of non-basal slips (pyramidal slip or prismatic slip). The activation of non-basal slips would delay the texture strengthening and then release the stress concentration. Koike et al. pointed out that non-basal slips could be activated when the grain size was less than 10 μm, and under this case, the cross-slip of <a> dislocation from the basal to the prismatic plane occurred readily [35]. Similarly, Zhao et al. and Mayama et al. found that the profuse prismatic <a> dislocation activations would suppress twinning formation by rotating grains to preferred orientations for further deformation, resulting in improved ductility [36,37]. In our recent work, a considerable proportion of prismatic <a> dislocations was activated during the tension deformation of AZ31 magnesium alloys, which was closely related to their higher tensile elongation [38]. The resulting sheet possessed a finer grain size less than 10 μm, which was conducive the activation of non-basal slips. The evidence could be obtained by the TEM characteristics of the necking part of the tension sample (Figure 11). Figure 11a,b depicts abundant dislocation pile-up and twins, respectively. In such a complex dislocation tangle, non-basal dislocations are bound to participate. Therefore, the activation of the non-basal slip produced by a finer grain size partly contributed to the higher fracture elongation.

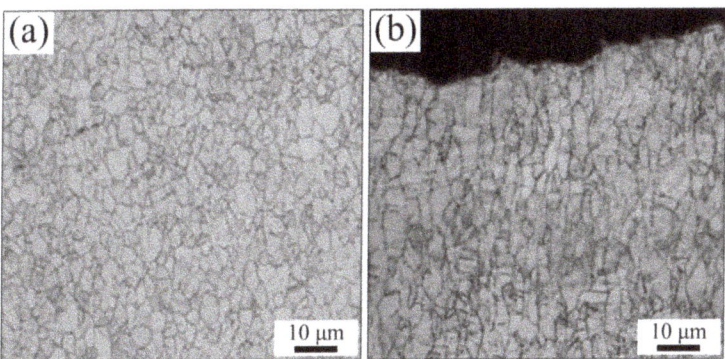

Figure 10. Metallographic observations of the uniform deformation part A (**a**) and necking part B (**b**) indicated by the black point in Figure 11a.

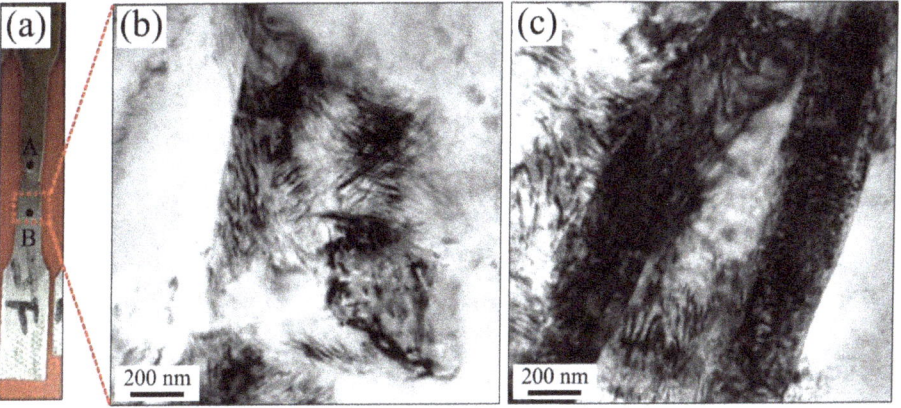

Figure 11. TEM characteristics of the necking part of the AZ31 magnesium alloy sheets after tension deformation: (**a**) tension sample; (**b**) dislocation pile-up; (**c**) twins.

Figure 12 shows the fracture SEM characteristics of the AZ31 tension samples fabricated by multi-pass lowered-temperature rolling. Clearly, the fracture of the as-cast AZ31 tension sample exhibited a typical quasi-cleavage fracture appearance with a smooth cleavage plane and cleavage step (Figure 12a). Similarly, the tension samples after one pass and two passes also had a quasi-cleavage fracture appearance. These features suggested that the above tension samples should undergo brittle failure. However, in Passes 3–5, abundant tear ridges and dimples can be seen in Figure 12d–f. This is generally named the brittle–ductile fracture mode, which is common in magnesium alloy sheets. As the rolling proceeded (a decrease in the grain size), the fracture mode gradually changed from brittle–ductile fracture to ductile fracture. Finally, because of the smaller grain size (2.5 µm) in the Pass 8 sheet, the tension fracture exhibited typical ductile fracture with dimples.

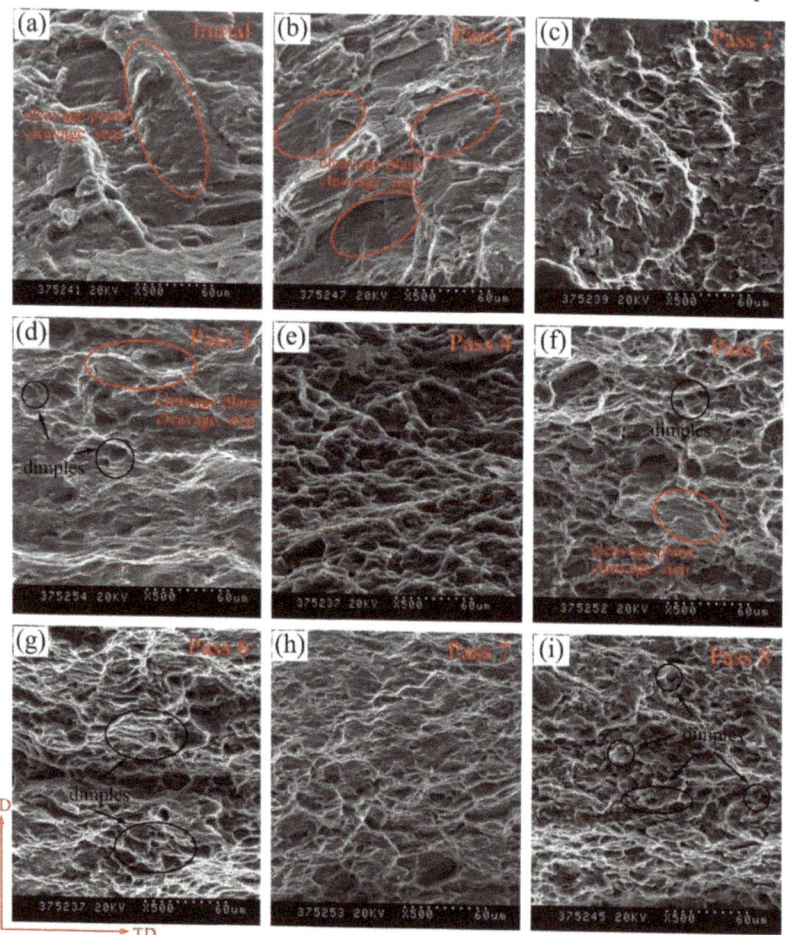

Figure 12. Fracture SEM characteristics of AZ31 tension samples fabricated by multi-pass lowered-temperature rolling: (**a**) Initial; (**b**) Pass 1; (**c**) Pass 2; (**d**) Pass 3; (**e**) Pass 4; (**f**) Pass 5; (**g**) Pass 6; (**h**) Pass 7 and (**i**) Pass 8.

4. Conclusions

In this work, multi-pass lowered-temperature rolling was applied to fabricate AZ31 magnesium alloy sheets with high mechanical properties. This rolling process not only realized the fabrication of thin sheets but also optimized the microstructure. Due to the effect of multiple dynamic recrystallization, the grain size was obviously refined and the microstructure homogeneity was distinctly improved. The average grain size in the resulting sheet was refined to 4.38 µm. As the rolling proceeded, the (0002) basal plane gradually rotated toward the rolling plane, but the participation of non-basal slip made the (0002) basal plane point slightly toward the RD. These microstructure variations influenced the yield stress greatly. In the resulting sheet, the yield stresses were markedly increased to 202 MPa along the TD and 189.8 MPa along the RD. In addition, the finer grain size was beneficial to the grain boundary sliding and the activation of non-basal slip, which finally improved the fracture elongations. In the resulting sheet, the fracture elongations reached 38% in the TD a nd 39.2% in the RD.

Author Contributions: Conceptualization, W.W. and W.C.; methodology, L.Z., W.W., Z.W., B.S., W.C., Y.Y. and W.Z.; software, Z.W., L.Z., Z.W., B.S., Y.Y. and W.Z.; validation, L.Z., Z.W. and W.C.; formal analysis, B.S. and Y.Y.; investigation, Q.M., W.W. and W.C.; resources, W.C. and W.Z.; data curation, Q.M. and L.Z.; writing—original draft preparation, Q.M.; writing—review and editing, W.W.; visualization, L.Z., Z.W., B.S. and Y.Y.; supervision, B.S., Y.Y. and W.Z.; funding acquisition, W.W., W.C. and W.Z. All authors have read and agreed to the published version of the manuscript.

Funding: This work was funded by the National Natural Science Foundation of China (Grant No. 51975146, 52205344), the Natural Science Foundation of Shandong Province (Grant No. ZR2020QE171, ZR2021ME073), and the Key Research and Development Plan in Shandong Province (Grant No. 2019JZZY010364).

Data Availability Statement: Research data are available upon reasonable request to the authors.

Conflicts of Interest: The authors declare no conflict of interest.

References

1. Zhang, L.; Zhang, J.; Leng, Z.; Liu, S.; Yang, Q.; Wu, R.; Zhang, M. Microstructure and mechanical properties of high-performance Mg-Y-Er-Zn extruded alloy. *Mater. Des.* **2014**, *54*, 256–263. [CrossRef]
2. Da Huo, P.; Li, F.; Wang, Y.; Wu, R.Z.; Gao, R.H.; Zhang, A.X. Annealing coordinates the deformation of shear band to improve the microstructure difference and simultaneously promote the strength-plasticity of composite plate. *Mater. Des.* **2022**, *219*, 110696. [CrossRef]
3. Suh, B.-C.; Shim, M.; Shin, K.S.; Kim, N.J. Current issues in magnesium sheet alloys: Where do we go from here? *Scr. Mater.* **2014**, *84–85*, 1–6. [CrossRef]
4. Chino, Y.; Sassa, K.; Kamiya, A.; Mabuchi, M. Stretch formability at elevated temperature of a cross-rolled AZ31 Mg alloy sheet with different rolling routes. *Mater. Sci. Eng. A* **2008**, *473*, 195–200. [CrossRef]
5. Wang, D.; Liu, S.; Wu, R.; Zhang, S.; Wang, Y.; Wu, H.; Zhang, J.; Hou, L. Synergistically improved damping, elastic modulus and mechanical properties of rolled Mg-8Li-4Y-2Er-2Zn-0.6Zr alloy with twins and longperiod stacking ordered phase. *J. Alloy. Compd.* **2021**, *881*, 160663. [CrossRef]
6. Chen, W.Z.; Yu, Y.; Wang, X.; Wang, E.; Liu, Z. Optimization of rolling temperature for ZK61 alloy sheets via microstructure uniformity analysis. *Mater. Sci. Eng. A* **2013**, *575*, 136–143. [CrossRef]
7. Huang, X.; Chino, Y.; Mabuchi, M.; Matsuda, M. Influences of grain size on mechanical properties and cold formability of Mg-3Al-1Zn alloy sheets with similar weak initial textures. *Mater. Sci. Eng. A* **2014**, *611*, 152–161. [CrossRef]
8. Huang, X.; Suzuki, K.; Chino, Y.; Mabuchi, M. Texture and stretch formability of AZ61 and AM60 magnesium alloy sheets processed by high-temperature rolling. *J. Alloy. Compd.* **2015**, *632*, 94–102. [CrossRef]
9. Huo, Q.; Yang, X.; Sun, H.; Li, B.; Qin, J.; Wang, J.; Ma, J. Enhancement of tensile ductility and stretch formability of AZ31 magnesium alloy sheet processed by cross-wavy bending. *J. Alloy. Compd.* **2013**, *581*, 230–235. [CrossRef]
10. Wang, W.; Chen, W.; Zhang, W.; Cui, G.; Wang, E. Effect of deformation temperature on texture and mechanical properties of ZK60 magnesium alloy sheet rolled by multi-pass lowered-temperature rolling. *Mater. Sci. Eng. A* **2018**, *712*, 608–615. [CrossRef]
11. Cho, J.-H.; Jeong, S.S.; Kang, S.-B. Deep drawing of ZK60 magnesium sheets fabricated using ingot and twin-roll casting methods. *Mater. Des.* **2016**, *110*, 214–224. [CrossRef]
12. Kim, W.; Yoo, S.; Chen, Z.; Jeong, H. Grain size and texture control of Mg-3Al-1Zn alloy sheet using a combination of equal-channel angular rolling and high-speed-ratio differential speed-rolling processes. *Scr. Mater.* **2009**, *60*, 897–900. [CrossRef]
13. Suh, J.; Victoria-Hernández, J.; Letzig, D.; Golle, R.; Volk, W. Enhanced mechanical behavior and reduced mechanical anisotropy of AZ31 Mg alloy sheet processed by ECAP. *Mater. Sci. Eng. A* **2016**, *650*, 523–529. [CrossRef]
14. Zhang, L.; Chen, W.; Zhang, W.; Wang, W.; Wang, E. Microstructure and mechanical properties of thin ZK61 magnesium alloy sheets by extrusion and multi-pass rolling with lowered temperature. *J. Mater. Process. Technol.* **2016**, *237*, 65–74. [CrossRef]
15. Wang, W.; Zhang, W.; Chen, W.; Yang, J.; Zhang, L.; Wang, E. Homogeneity improvement of friction stir welded ZK61 alloy sheets in microstructure and mechanical properties by multi-pass lowered-temperature rolling. *Mater. Sci. Eng. A* **2017**, *703*, 17–26. [CrossRef]
16. Chen, W.Z.; Zhang, W.C.; Zhang, L.X.; Wang, E.D. Property improvements in fine-grained Mg-Zn-Zr alloy sheets produced by temperature-step-down multi-pass rolling. *J. Alloy. Compd.* **2015**, *646*, 195–203. [CrossRef]
17. Prakash, P.; Toscano, D.; Shaha, S.K.; Wells, M.A.; Jahed, H.; Williams, B.W. Effect of temperature on the hot deformation behavior of AZ80 magnesium alloy. *Mater. Sci. Eng. A* **2020**, *794*, 139923. [CrossRef]
18. Zhou, T.; Yang, Z.; Hu, D.; Feng, T.; Yang, M.; Zhai, X. Effect of the final rolling speeds on the stretch formability of AZ31 alloy sheet rolled at a high temperature. *J. Alloy. Compd.* **2015**, *650*, 436–443. [CrossRef]
19. Chen, W.; Wang, X.; Hu, L.; Wang, E. Fabrication of ZK60 magnesium alloy thin sheets with improved ductility by cold rolling and annealing treatment. *Mater. Des.* **2012**, *40*, 319–323. [CrossRef]
20. Young, J.P.; Ayoub, G.; Mansoor, B.; Field, D.P. The effect of hot rolling on the microstructure, texture and mechanical properties of twin roll cast AZ31Mg. *J. Mater. Process. Technol.* **2015**, *216*, 315–327. [CrossRef]

21. Meher, A.; Mahapatra, M.M.; Samal, P.; Vundavilli, P.R.; Shankar, K.V. Statistical Modeling of the Machinability of an In-Situ Synthesized RZ5/TiB$_2$ Magnesium Matrix Composite in Dry Turning Condition. *Crystals* **2022**, *12*, 1353. [CrossRef]
22. Wang, W.; Cui, G.; Zhang, W.; Chen, W.; Wang, E. Evolution of microstructure, texture and mechanical properties of ZK60 magnesium alloy in a single rolling pass. *Mater. Sci. Eng. A* **2018**, *724*, 486–492. [CrossRef]
23. Imandoust, A.; Barrett, C.D.; Oppedal, A.L.; Whittington, W.R.; Paudel, Y.; El Kadiri, H. Nucleation and preferential growth mechanism of recrystallization texture in high purity binary magnesium-rare earth alloys. *Acta Mater.* **2017**, *138*, 27–41. [CrossRef]
24. Chen, W.Z.; Wang, X.; Kyalo, M.N.; Wang, E.D.; Liu, Z. Yield strength behavior for rolled magnesium alloy sheets with texture variation. *Mater. Sci. Eng. A* **2013**, *580*, 77–82. [CrossRef]
25. Trang, T.T.T.; Zhang, J.H.; Kim, J.H.; Zargaran, A.; Hwang, J.H.; Suh, B.-C.; Kim, N.J. Designing a magnesium alloy with high strength and high formability. *Nat. Commun.* **2018**, *9*, 2522. [CrossRef]
26. Wang, Y.N.; Huang, J.C. Texture analysis in hexagonal materials. *Mater. Chem. Phys.* **2003**, *81*, 11–26. [CrossRef]
27. Li, G.; Zhang, J.; Wu, R.; Feng, Y.; Liu, S.; Wang, X.; Jiao, Y.; Yang, Q.; Meng, J. Development of high mechanical properties and moderate thermal conductivity cast Mg alloy with multiple RE via heat treatment. *J. Mater. Sci. Technol.* **2017**, *34*, 1076–1084. [CrossRef]
28. Wang, W.; Chen, W.; Zhang, W.; Cui, G.; Wang, E. Weakened anisotropy of mechanical properties in rolled ZK60 magnesium alloy sheets with elevated deformation temperature. *J. Mater. Sci. Technol.* **2018**, *34*, 2042–2050. [CrossRef]
29. Liu, D.; Liu, Z.; Wang, E. Effect of rolling reduction on microstructure, texture, mechanical properties and mechanical anisotropy of AZ31 magnesium alloys. *Mater. Sci. Eng. A* **2014**, *612*, 208–213. [CrossRef]
30. Liu, D.; Bian, M.Z.; Zhu, S.; Chen, W.Z.; Liu, Z.; Wang, E.D.; Nie, J. Microstructure and tensile properties of Mg-3Al-1Zn sheets produced by hot-roller-cold-material rolling. *Mater. Sci. Eng. A* **2017**, *706*, 304–310. [CrossRef]
31. Koike, J. Enhanced deformation mechanisms by anisotropic plasticity in polycrystalline Mg alloys at room temperature. *Met. Mater. Trans. A* **2005**, *36*, 1689–1696. [CrossRef]
32. Yoshida, Y.; Arai, K.; Itoh, S.; Kamado, S.; Kojima, Y. Realization of high strength and high ductility for AZ61 magnesium alloy by severe warm working. *Sci. Technol. Adv. Mater.* **2005**, *6*, 185–194. [CrossRef]
33. Koike, J.; Ohyama, R.; Kobayashi, T.; Suzuki, M.; Maruyama, K. Grain-Boundary Sliding in AZ31 Magnesium Alloys at Room Temperature to 523 K. *Mater. Trans.* **2003**, *44*, 445–451. [CrossRef]
34. Zheng, R.; Du, J.-P.; Gao, S.; Somekawa, H.; Ogata, S.; Tsuji, N. Transition of dominant deformation mode in bulk polycrystalline pure Mg by ultra-grain refinement down to sub-micrometer. *Acta Mater.* **2020**, *198*, 35–46. [CrossRef]
35. Koike, J.; Kobayashi, T.; Mukai, T.; Watanabe, H.; Suzuki, M.; Maruyama, K.; Higashi, K. The activity of non-basal slip systems and dynamic recovery at room temperature in fine-grained AZ31B magnesium alloys. *Acta Mater.* **2003**, *51*, 2055–2065. [CrossRef]
36. Zhao, D.; Ma, X.; Srivastava, A.; Turner, G.; Karaman, I.; Xie, K.Y. Significant disparity of non-basal dislocation activities in hot-rolled highly-textured Mg and Mg-3Al-1Zn alloy under tension. *Acta Mater.* **2021**, *207*, 116691. [CrossRef]
37. Mayama, T.; Noda, M.; Chiba, R.; Kuroda, M. Crystal plasticity analysis of texture development in magnesium alloy during extrusion. *Int. J. Plast.* **2011**, *27*, 1916–1935. [CrossRef]
38. Zhang, Z.; Zhang, J.; Wang, W.; Liu, S.; Sun, B.; Xie, J.; Xiao, T. Unveiling the deformation mechanism of highly deformable magnesium alloy with heterogeneous grains. *Scr. Mater.* **2022**, *221*, 114963. [CrossRef]

Article

High Strain Rate Deformation Behavior of Gradient Rolling AZ31 Alloys

Yingjie Li [1], Hui Yu [1,*], Chao Liu [1], Yu Liu [2], Wei Yu [3], Yuling Xu [4], Binan Jiang [5], Kwangseon Shin [6] and Fuxing Yin [7,*]

1. Tianjin Key Laboratory of Materials Laminating Fabrication and Interfacial Controlling Technology, School of Materials Science and Engineering, Hebei University of Technology, Tianjin 300130, China; a2323496328@163.com (Y.L.); sk8liuchao@163.com (C.L.)
2. School of Materials Science and Engineering, Hunan University, Changsha 410082, China; lyahaqhn@163.com
3. School of Materials Science and Engineering, Hefei University of Technology, Hefei 200039, China; yuwei52213@163.com
4. Baosteel Metal Co., Ltd., Shanghai 200940, China; xuyuling@baosteel.com
5. PLA Army Academy of Artillery and Air Defense, Hefei 230031, China; tata_maxwell@163.com
6. Department of Materials Science and Engineering, Seoul National University, Seoul 08826, Republic of Korea; ksshin@snu.ac.kr
7. Institute of New Materials, Guangdong Academy of Science, Guangzhou 510651, China
* Correspondence: huiyu@vip.126.com (H.Y.); yinfuxing@hebut.edu.cn (F.Y.)

Abstract: A dynamic impact test was performed on as-rolled AZ31 alloys with gradient microstructure under various strains. The microstructural evolution and mechanical properties were systematically investigated. As the strain rate gradually increased, an increasing number of twins were formed, facilitating dynamic recrystallization (DRX), and the mechanical properties were also gradually improved. The microstructure became heterogeneous at higher strain rates, but the peak stress decreased. The impact process resulted in a significantly higher performance due to microstructural refinement, work hardening by dislocations, and precipitates. In addition, both the adiabatic shear band and the adjacent crack experienced a temperature rise that exceeded the recrystallization temperature of the alloys. This observation also explains the presence of ultrafine recrystallized grains within the adiabatic shear band and the appearance of molten metal around the crack.

Keywords: AZ31; high strain rate; gradient rolling; microstructure; mechanical property

1. Introduction

Magnesium (Mg) is recognized as the fourth most abundant metal worldwide. Its lightweight nature, combined with its enhanced strength and processing capabilities, has led to the widespread utilization of Mg alloys across various industries such as aviation, automobiles, electronics, and medicine [1–5]. Particularly in the case of aerospace and automotive applications, magnesium alloys often experience dynamic loading, including impact, collision, and explosion, which necessitates the need to characterize and improve the mechanical properties of magnesium alloys at high strain rates. Typically, wrought Mg alloys manufactured by extrusion and rolling processes demonstrate superior properties [6–10], and they also result in a distinct fiber texture, leading to noticeable anisotropy [11–14]. However, the experimental samples are small in size, and only a single microstructure can generally be formed after plastic deformation, while the actual magnesium alloy forgings lack a uniform microstructure due to the different degrees of deformation in each part. Gradient rolling can produce samples with different deformation degrees at the same time, thus greatly reducing the test cost and experimental error [15].

Recently, the majority of research efforts on the deformation behavior of Mg alloys have primarily focused on quasi-static conditions, with fewer investigations into high strain rate deformation behavior [16]. In general, the deformation mechanism may be altered under high strain rate compression compared to quasi-static compression [17–19]. Thus, exploring the microstructural evolution of Mg alloys under high strain rates would play a crucial role in understanding such a phenomenon, which, in turn, would allow for a better understanding of related mechanical properties [20–25]. For instance, Yu et al. [26] conducted dynamic impact tests on the EW75 Mg alloy, demonstrating that an increase in strain rate correspondingly led to a higher number of twins and recrystallized grains, which resulted in improved mechanical properties. Moreover, Ji et al. [27] observed a diminishing basal texture trend in the edge and central regions of a Mg alloy using the cross-rolling process.

However, the existing literature primarily focuses on individual processing and lacks systematic experiments on the high strain rate deformation behavior of magnesium alloys. Therefore, this study aims to bridge this research gap by conducting dynamic impact tests on AZ31 Mg alloys. All specimens underwent varying levels of deformation and were exposed to different strain rates. The main objective was to investigate how high strain rates influence the mechanical properties and microstructural evolution of gradient-deformed Mg alloys. The findings will not only contribute to a theoretical foundation but will also provide technical insight for developing high performance Mg alloys used in dynamic load conditions.

2. Experimental Section

A commercial AZ31 (Mg-3.25Al-0.92Zn-0.34Mn, wt.%) Mg alloy (provided by Dongguan Kuangyu Metal Materials Co., Ltd., Dongguan, China) with a diameter of 60 mm and a height of 120 mm was used. The specimen was extruded and deformed into a cylindrical bar with a diameter of 20 mm, then turned into a conical workpiece with a diameter of 10 mm at one end and a 20 mm diameter at the other end, and then rolled at 723 K through a hole pattern in order to obtain transition tissues with different strain levels. The methodology employed to achieve a gradient structure was comprehensively described in our previous study [28]. Dynamic impact tests were conducted on samples with rolling reductions of 0%, 10%, 20%, and 30% (referred to as R0, R10, R20, and R30, respectively). A high-speed impact specimen shape of Ø8 mm × 5 mm cylindrical specimen was used. These tests were performed utilizing a split Hopkinson bar testing setup (ARCHIMEDESALT 1000, ARCHIMEDES, Tianjin, China), with impact pressures set at 0.15 MPa, 0.2 MPa, 0.25 MPa, and 0.3 MPa, corresponding to strain rates of $800~s^{-1}$, $1400~s^{-1}$, $2000~s^{-1}$, and $2400~s^{-1}$, respectively.

For microstructural analysis, an optical microscope (OM, OLYCIA M3) and a scanning electron microscope (SEM, JSM-6510A, JEOL Ltd., Tokyo, Japan) were employed. The Φ3 mm disk foils were carefully prepared by grinding and punching. Subsequently, the samples underwent electrolytic jet polishing and ion milling. Transmission electron microscope (TEM) analysis was conducted using a JEM2100F with an energy dispersive X-ray spectrometer (EDAX-TSL, JEOL Ltd., Tokyo, Japan). The electron backscatter diffraction (EBSD) samples were subjected to argon ion polishing using the Hitachi Implus400 system (Hitachi, Tokyo, Japan) and were characterized using an SEM equipped with an Oxford C-nanoprobe (Zeiss Gemini 300, Carl Zeiss, Jena, Germany). To ensure accurate acquisition of EBSD datasets, a scanning step of 0.8 mm was implemented.

3. Results and Discussion

3.1. Microstructural Evolution

Figure 1a–d shows the OM of the four strain samples before impact, and Figure 1e–t illustrates the OM of the four AZ31 samples following impact at various strain rates. The evolution of the impacted samples can be categorized into four stages as the strain rate increased. (1) At a strain rate of $800~s^{-1}$, a significant number of twins formed across all samples. (2) At a strain rate of $1400~s^{-1}$, the previously formed deformation twins largely disappeared, and further impact led to grain size reduction. (3) At a strain rate of $2000~s^{-1}$, the impacted specimen experienced further growth in grain size accompanied by a noticeable rise in temperature. (4) At a strain rate of 2400 s^{-1}, all samples underwent fracture, characterized by a prominent presence of adiabatic shear bands and deformation twins within the sample. Clearly, the extent of temperature rise was positively correlated with the magnitude of the strain rate [16,29,30], which, in turn, allowed for the coarsening of the grain size to be imaged.

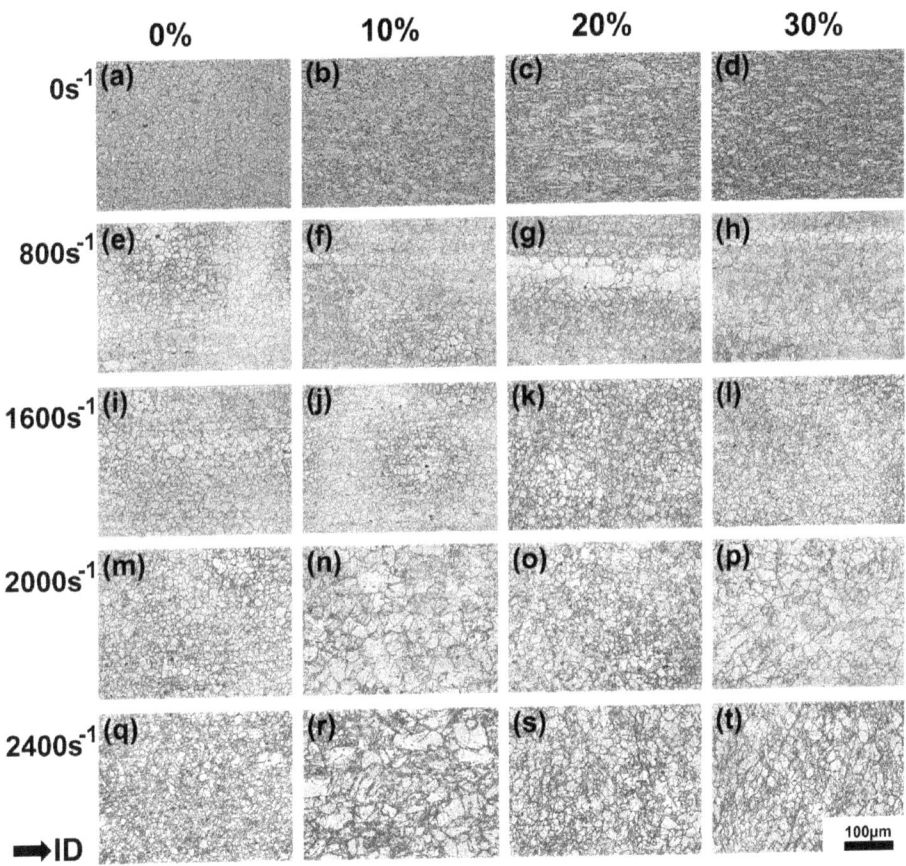

Figure 1. OM of AZ31 alloys with different rolling reductions of 0%, 10%, 20%, and 30% under different strain rates: (**a–d**) initial state, (**e–h**) 800 s^{-1}, (**i–l**) 1600 s^{-1}, (**m–p**) 2000 s^{-1}, and (**q–t**) 2400 s^{-1}. ID: impact direction.

Figure 2a–d presents the inverse pole figure (IPF) and grain size distribution diagram of samples that underwent different rolling reductions after impact at a strain rate of 2000 s^{-1}. The impact-induced grain structure of the R0 and R30 samples appeared relatively uniform, while the R10 and R20 samples exhibited significant variation in grain size. In particular, the maximum grain size observed in the R10 sample was approximately 14.5 times larger than the average grain size (AGS). This discrepancy can be attributed to two factors. On one hand, the short impact deformation time resulted in some grains failing to promptly coordinate deformation, leading to significant disparities in grain distribution. On the other hand, the initial grain homogeneity also influenced this phenomenon. It is important to highlight that irrespective of the AGS before impact deformation, the AGS of each sample converged to around 9 μm after impact at a strain rate of 2000 s^{-1}. Even when the morphology varied considerably among all samples, the impact-induced deformation facilitated the formation of numerous fine grains within the coarse grains, enabling coordinated deformation and a remarkable reduction in AGS. These findings align with observations from previous dynamic impact studies on Mg alloys [31–33].

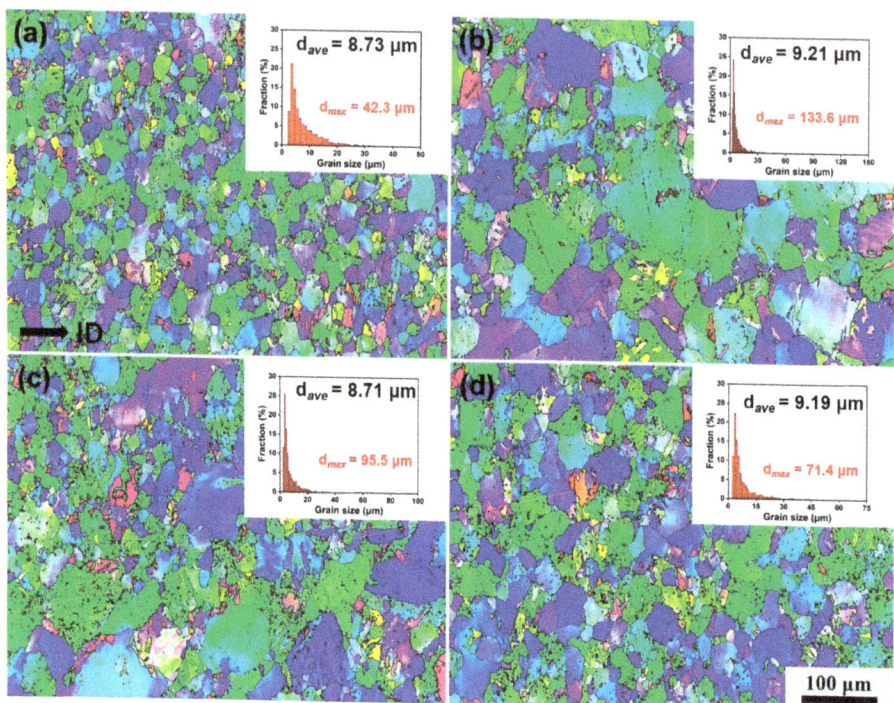

Figure 2. IPF and grain size distribution of AZ31 alloys with various rolling reductions after impact at a strain rate of 2000 s^{-1}: (**a**) 0%; (**b**) 10%; (**c**) 20%; and (**d**) 30%. ID: impact direction.

Figure 3 shows the texture distribution of the samples after impact at a strain rate of 2000 s^{-1}. There were no significant variations in texture intensity. The (0001) pole figure (PF) demonstrated a predominantly parallel distribution in relation to the impact direction (ID). Additionally, an IPF revealed the formation of a non-fiber texture component that was roughly parallel to the transverse direction (TD). In addition, the texture components were also detected in transitional orientations between the [10-10]||TD and [11-20]||TD.

When the Mg alloy was impacted at high speed, a considerable number of deformed grains were formed to accommodate the severe deformation [34]. Figure 4 presents the distribution and proportion of grains below the AGS in the impacted samples, along with their corresponding textures. The volume fraction (V_f) of grains below the AGS was approximately 13%, and a highly uneven grain distribution was found due to more unbroken coarse grains when compared with Figure 2. Wang et al. [35] observed that when the compression direction (CD) aligns with the rolling direction (RD), the texture is enhanced. Conversely, when the CD is perpendicular to the RD, the texture becomes weaker.

The corresponding (0001) PF revealed that while certain grains maintained an orientation in which the c-axis was parallel to the impact direction (ID), most grains tended to diffuse along the normal direction (ND) and formed orientations offset by about 30° in the transverse direction (TD). The color distribution in the figure shows that the texture formed after impact became more pronounced as the rolling reduction increased. Weak texture components were observed in the R0 and R10 samples, while the R20 and R30 specimens exhibited distinct base poles. This illustrates that when the two deformations shared the same direction, the deformation during the initial process also influenced the texture generated by the subsequent deformation.

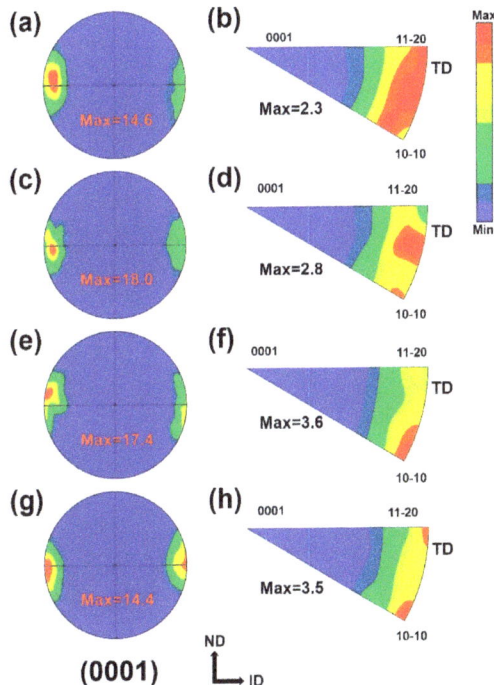

Figure 3. (**a,c,e,g**) (0001) PF and (**b,d,f,h**) IPF of specimens with different reductions (0%, 10%, 20%, and 30%) after impact at a strain rate of 2000 s^{-1}, respectively.

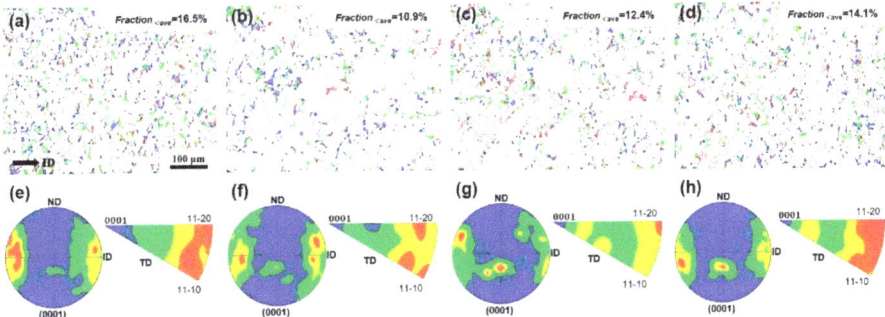

Figure 4. (**a–d**) IPF and proportion of the area with grain size below AGS after impact of the samples with deformation of 0%, 10%, 20%, and 30% at a strain rate of 2000 s^{-1} and (**e–h**) corresponding texture.

Generally, the deformation twins observed at high strain rates directly contribute to an increase in the average orientation difference (known as KAM), as depicted in Figure 5. The impacted samples displayed remarkably high KAM values, and the distribution of KAM within the coarse grains exhibited notable heterogeneity [36,37]. This phenomenon can be attributed to two factors: (1) The very short impact process limited the ability of some coarse grains to promptly undergo coordinated deformation. As a result, a nonuniform strain emerged within these grains. (2) A multitude of small grains involving dynamic recrystallization (DRX) formed within the coarse grains subsequent to impact. These newly formed grains exhibited relatively low KAM values and actively contributed to coordinating and influencing the KAM distribution to a significant extent. A similar finding was observed by Deng et al. [38], who proposed that twinning promotes the generation of slip to further achieve coordinated deformation. Thus, a substantial number of twins were indeed produced during the impact process to facilitate coordinated deformation, resulting in a reduction in the AGS.

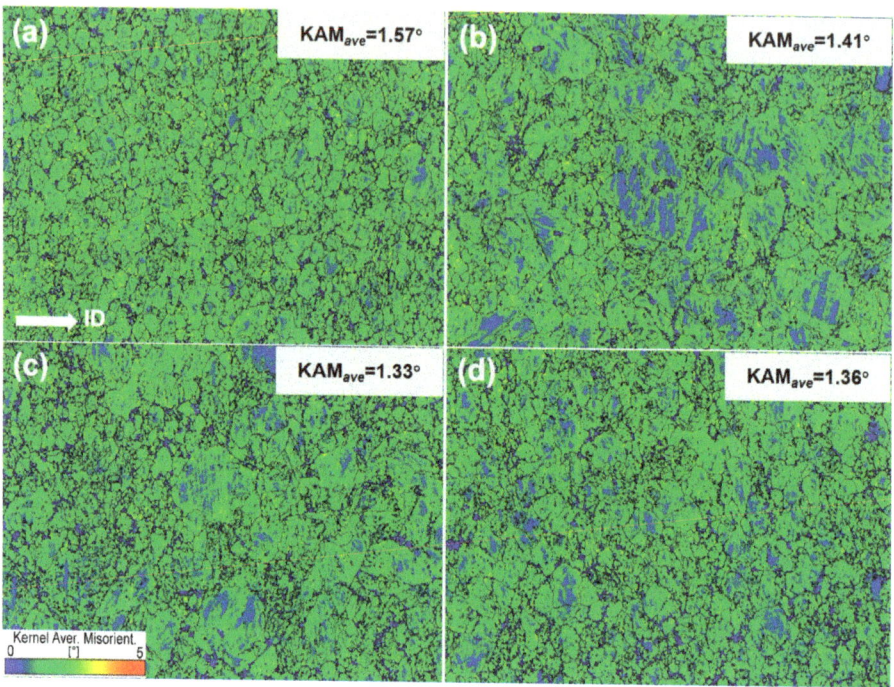

Figure 5. KAM distribution and average KAM values of samples with various rolling reductions after impact at a strain rate of 2000 s^{-1}: (**a**) 0%; (**b**) 10%; (**c**) 20%; and (**d**) 30%.

In addition, the microstructure of the samples after impact predominantly contained tensile twins, while compression twins and double twins were nearly absent. These tensile twins were primarily observed within the coarse grains, which highlights that tensile twins dominated the initial twinning deformation at high strain rates, as suggested by Chen et al. [39]. Figure 6 illustrates the specific distribution and V_f of the twins in specimen. For instance, in the case of the impact samples at a strain rate of 2000 s^{-1}, the R20 specimen exhibited the highest number of tensile twins, with a V_f of 23.4%. The formation of tensile twins mainly arose from certain original grains with basal texture. Additionally, due to the low critical resolved shear stress (CRSS) value associated with activating tensile twins, the strains corresponding to the impact deformation stage were fully sufficient to induce the highest number of tensile twins.

Comparing the IPF in Figure 2, nearly all the tensile twins form within the grains tended to exhibit a distribution aligned with the [11-20]||TD, as illustrated in Figure 4. However, the actual orientation of the twins exhibited a noticeable deviation from this direction. As a result, the texture with a higher quantity of formed twins after impact significantly weakened in the [11-20]||TD. Furthermore, it is notable that in Figure 4e, no non-fiber texture parallel to the TD formed. Gao et al. [40] also discovered that twins not only reduce the intensity of the texture but also alter the type of texture. In addition, the formation of numerous tensile twins within specific coarse grains largely affected the neighboring grain orientations. To conduct a comprehensive analysis, we focused on grains within two typical designated green rectangular frames, as depicted in Figure 6. The details are presented in Figure 7, using "P" to denote the parent grain and "ETs" to represent the tensile twins.

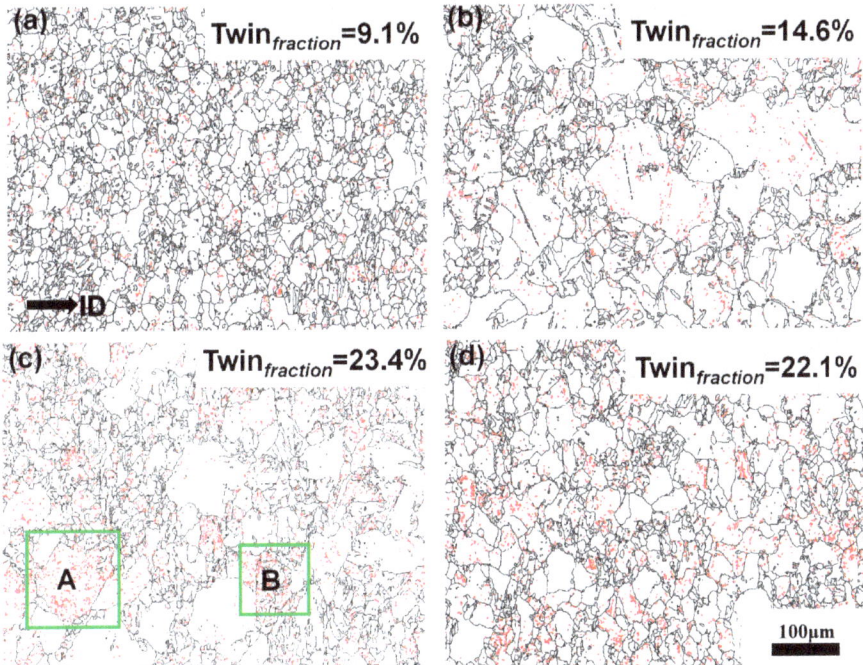

Figure 6. Twin distribution and its fraction with different rolling reductions after impact at a strain rate of 2000 s^{-1}: (**a**) 0%; (**b**) 10%; (**c**) 20%; and (**d**) 30%.

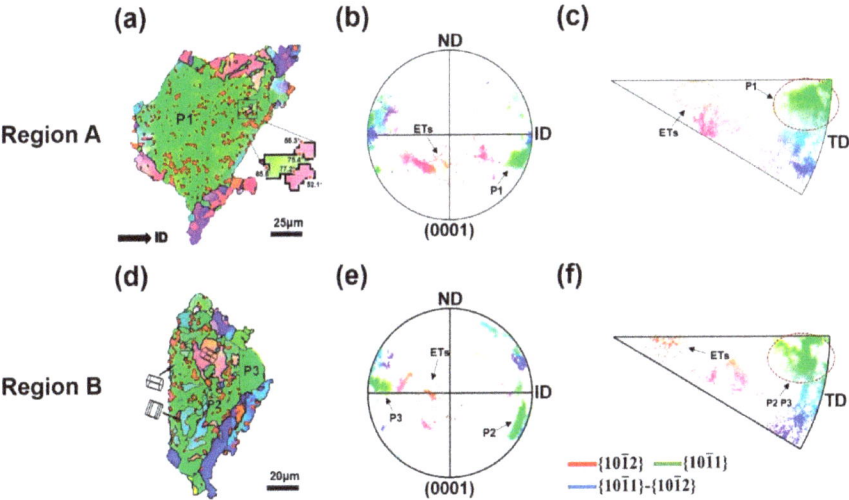

Figure 7. The EBSD result of the green rectangular box region in Figure 6 (**c**): (**a**,**d**) enlarged grains, and (**b**,**e**) and (**c**,**f**) correspond to (0001) PF and IPF, respectively.

In region A, a considerable number of tensile twins were observed within the P1 grains, and their orientations were generally consistent with the grains below the AGS. The corresponding texture components were formed near a 30° deviation from the ND to the TD in the (0001) PF. Conversely, the orientations of the grains surrounding P1 differed significantly from P1 itself. Some grains exhibited texture components similar to those of the tensile twins, while others aligned with the [10-10]||TD. Particularly, the black rectangular frame in Figure 7a revealed compelling evidence of the influence of tensile twins on the orientation of surrounding grains. A discernible gradient trend was observed in the grain boundary (GB) orientation difference among the three small grains formed adjacent to the twins. Moreover, the colors of these grains underwent significant changes. These observations further emphasize that the formation of tensile twins substantially impacts the orientation changes of the surrounding grains [41,42].

In region B, a number of tensile twins were also formed within the P2 and P3 grains. However, unlike the twins in region A, only a few grains surrounding the parent grains of these twins exhibited similar orientations. The texture orientations of the majority of the grains aligned with the [10-10]||TD, with only a small number of grains sharing a similar orientation to their parent grains. Notably, it is important to highlight that grains with significantly different orientations formed at higher twin densities within the P2 grains. The texture orientations of the small grains formed around these grains closely resembled their own orientations, differing from the P2 grains. These observations indirectly indicate a notable influence of the formation of tensile twins on the texture components of the surrounding grains.

3.2. Mechanical Property

Figure 8 presents the true strain–stress curves of the gradient rolling samples with varying deformations after impact at different strain rates. All curves exhibited an S-shaped plot, which is characteristic of the twinning-dominated deformation mechanism in Mg alloys [43]. This kind of deformation process can be divided into three stages. Initially, during the early stage of impact, the alloy undergoes significant work hardening, with the stress rapidly increasing as strain accumulates. Once a certain threshold of strain is reached, a distinct yield platform becomes evident, and the stress exhibits a gradual upward trend. The presence of a low-yield platform arises from the loading direction induced by impact, favoring the formation of tensile twins in the initial stages of deformation. The relatively low CRSS required for tensile twin formation allows the material to adjust its deformation by generating such twins, leading to an earlier attainment of the yield condition. As deformation progresses further, the stress continues to rise, but the material experiences a competition between strain hardening and an adiabatic temperature rise associated with deformation at high strain rates. This competition results in fluctuations in the curve, leading to an oscillating behavior [44], and the stress reaches its peak value. Subsequently, the stress gradually decreases until deformation ceases. Figure 8e represents the true strain–stress curve under quasi-static compression. It can be observed that the shape of the curve is similar to those in Figure 8a–d, but the fluctuation region at the peak is absent, which indicates that there was no pronounced adiabatic temperature rise. Additionally, the peak stress was reduced, suggesting that higher strain rates contribute to work hardening.

Figure 9 illustrates the yield strength (YS) and peak compressive strength (CS) of all specimens in Figure 8. The R20 sample exhibited a higher YS compared to the other samples. Specifically, at a strain rate of 1400 s^{-1}, the YS of the R20 sample reached 175 MPa, surpassing the YS of the R0 sample at a strain rate of 0.001 s^{-1} by 65 MPa. In the case of the peak CS within the range from 0.001 s^{-1} to 2000 s^{-1}, the CS increased with higher strain rates. However, the impacted samples fractured when the strain rate reached 2400 s^{-1}, resulting in a significant decline in CS. Notably, the R20 sample consistently exhibited the highest CS across all strain rates. For example, the CS reached 644 MPa at a strain rate of 2000 s^{-1}, exceeding the CS of the same strain sample at a strain rate of 0.001 s^{-1} by 145 MPa. This also emphasizes the pronounced sensitivity of the CS of the impacted samples to the strain rate [30,45].

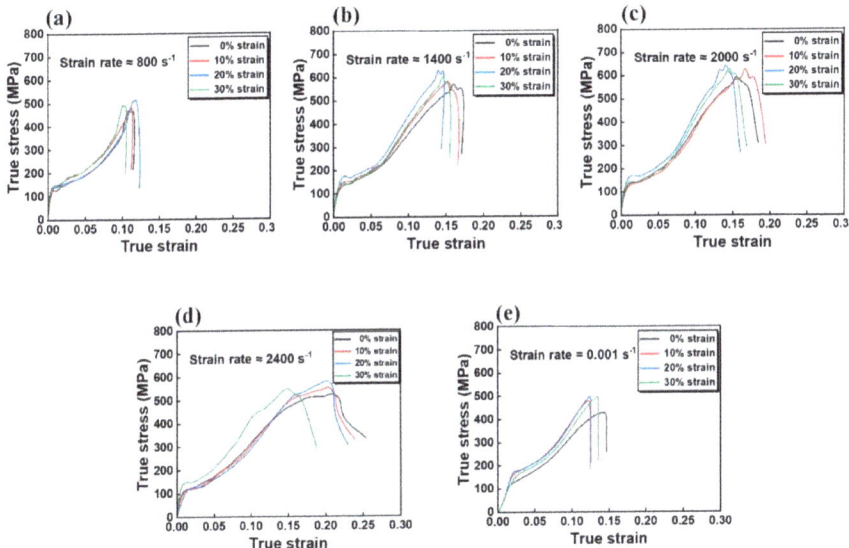

Figure 8. The true strain–stress curves of gradient rolling AZ31 alloys with different rolling reductions at strain rates of (**a**) 800 s^{-1}; (**b**) 1400 s^{-1}; (**c**) 2000 s^{-1}; (**d**) 2400 s^{-1}; and (**e**) 0.001 s^{-1}.

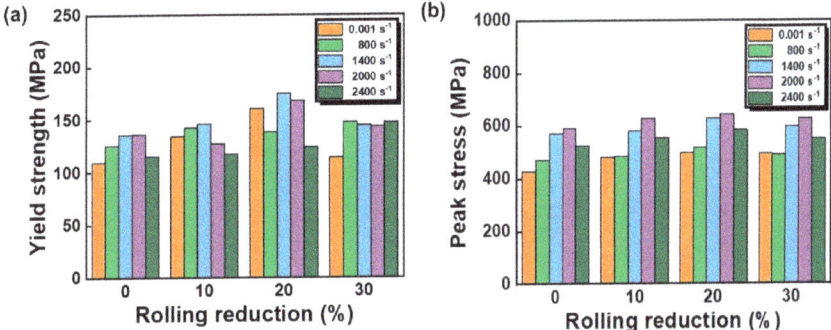

Figure 9. Summary of mechanical properties of all samples with different rolling reductions at various strain rates: (**a**) yield strength and (**b**) compressive strength.

Generally, the mechanical properties can be strengthened by solid solution strengthening, grain boundary strengthening, dislocation strengthening, and shear band strengthening [46–49]. TEM analysis was carried out for a better understanding of such strengthening mechanisms. Figure 10a displays the formation of high-density dislocations in the samples after impact (which agrees with Figure 5), demonstrating numerous dislocations at high strain rates using KAM analysis. Furthermore, the presence of layered structures in the dislocation region (see Figure 10b) and lattice distortions in the high-resolution image (see Figure 10c) confirm the existence of stacking faults (SFs) resulting from the interaction between SFs and a large number of dislocations. This interaction severely hindered the plastic deformation process, thereby improving the YS [50,51]. Additionally, a twin–twin interaction was observed within the impact specimen (see Figure 10d,e), which also contributed to strengthening of the AZ31 alloy. In addition to dislocations and twinning, Figure 10g reveals the presence of numerous nanoscale second-phase particles and some DRXed grains. Detailed EDS mapping indicated that these second particles were in the Al-Mn phases. Most of these particles were located near dislocations and hindered the movement of dislocations to some extent, thus enhancing these properties as well. In addition, in light of Figure 2, which exhibited lots of fine grains, a decrease in grain size by DRX enhanced the YS according to the well-known Hall–Petch equation. In the case of quasi-static compression at low strain rates, the YS and CS of as-rolled samples remained low. However, when the strain rate reached a medium level and the sample still did not fracture after impact, both the YS and CS increased significantly (i.e., R20 sample). When the strain rate reached a high enough level that it caused specimen failure, the corresponding strength decreased dramatically, and the alloy exhibited negative strain rate sensitivity [52].

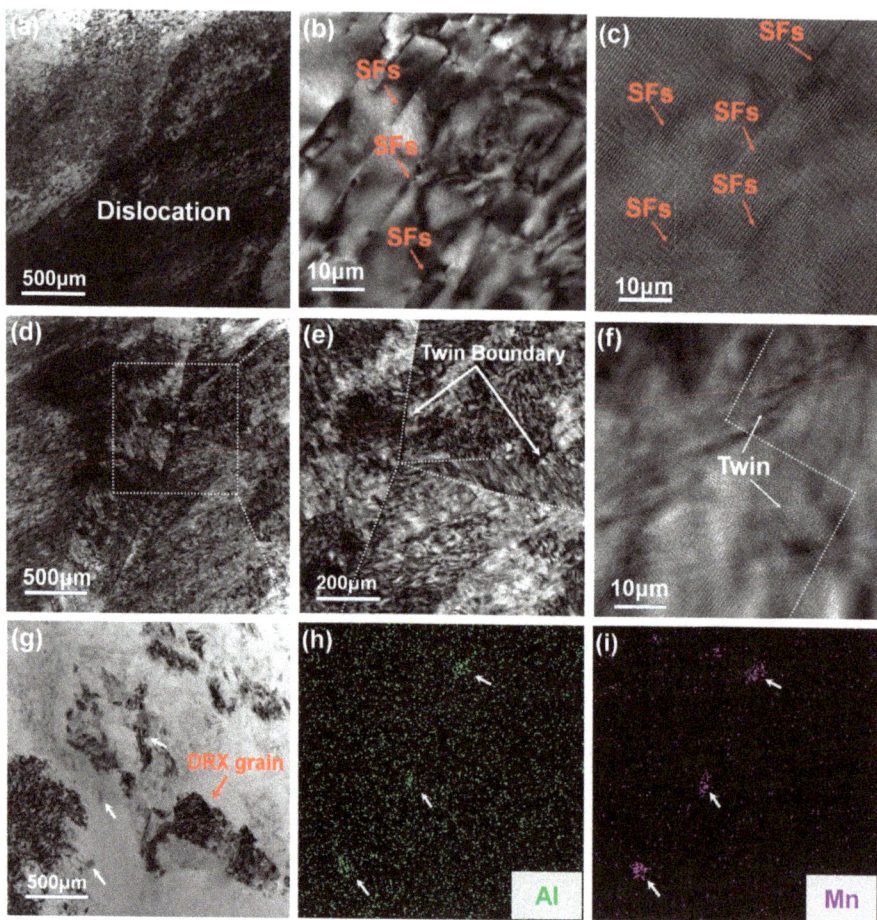

Figure 10. TEM of the sample with 20% rolling reduction after impact at a strain rate of 2000 s^{-1}: (a,b,d,e,g) bright-field image of defects and substructure; (c,f) high-resolution image corresponding to (b,e); and (h,i) EDS mapping of (g).

Figure 11 illustrates the Vickers hardness of AZ31 alloys subjected to different rolling reductions after impact. By examining the OM presented in Figure 1 and correlating it with these hardness values, it became apparent that the hardness values of all samples increased significantly with higher strain rates. However, at strain rates of 1400 s^{-1}, 2000 s^{-1}, and 2400 s^{-1}, the hardness values did not exhibit a positive strain rate sensitivity. Specifically, both the R0 and R30 samples consistently exhibited hardness values of approximately 70 HV after impact. In contrast, the hardness values of the R10 and R20 samples demonstrated a negative strain rate sensitivity with an increase in strain rate from 1400 s^{-1} to 2000 s^{-1}. This phenomenon was attributed to the significant increase in grain size observed in the

R10 and R20 samples following impact at a strain rate of 2000 s^{-1}. It is noteworthy that despite all samples fracturing at a strain rate of 2400 s^{-1}, the hardness measurement of the fractured sample indicates that the hardness value remained higher than the value prior to impact. This suggests that a substantial amount of deformation energy was absorbed by the sample prior to fracture/failure. Moreover, it is worth emphasizing that Figure 11 exhibits a significant hardness error bar for the R10 sample following impact at a strain rate of 2400 s^{-1}. Correlating this with the OM shown in Figure 1n, it becomes evident that the microstructure of the sample became highly heterogeneous after impact, primarily due to the presence of numerous shear bands. These measurements further support the well-established Hall–Petch relationship between hardness and grain size [53].

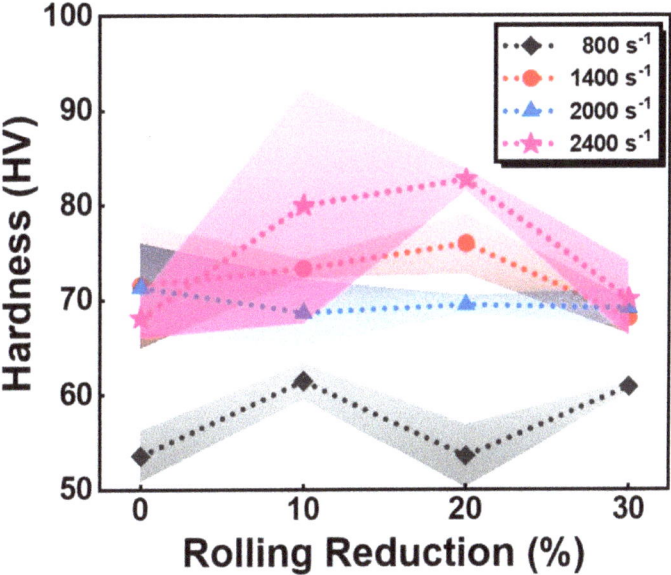

Figure 11. Vickers hardness of AZ31 alloys with different rolling reductions after impact at various strain rates.

Further examination of the shear band is presented in Figure 12. Although the grains within the shear band remained unevenly distributed, their size was significantly smaller compared to previous structures. Additionally, some nanocrystals and numerous nanoscale second phases in the shear band, as depicted in Figure 12b,d, contributed to the improved performance. However, at high strain rates, molten metal can be seen, as shown in Figure 12d (indicated by the red arrow), which, due to the local generation of heat during impact, resulted in limited heat dissipation and elevated temperatures within the shear band. Moreover, Figure 12b,d highlight the presence of micro-cracks surrounding specific grains, as indicated by the orange arrows. These micro-cracks have a tendency to propagate along adiabatic shear bands and can act as precursors to failure [54,55], providing an explanation for the fracture at the strain rate of 2400 s^{-1}.

In summary, when the as-rolled AZ30 alloy was subjected to deformation at high strain rates, a substantial number of deformation twins were generated, leading to an increase in dislocation density and noticeable work hardening. Concurrently, shear bands formed as dislocations slipped along specific GBs, resulting in the accumulation and interaction of dislocations, thereby enhancing the final performance. Moreover, due to the limited dissipation of heat, there was a rapid rise in temperature during short impact, leading to grain recrystallization, the occurrence of molten metal, and enhanced work hardening. All these factors contributed to a reduction in the AGS and an improvement in the mechanical properties. Additionally, the continued development of adiabatic shear bands was accompanied by the formation of cracks, which led to the failure and fragmentation of the alloys at excessively high strain rates.

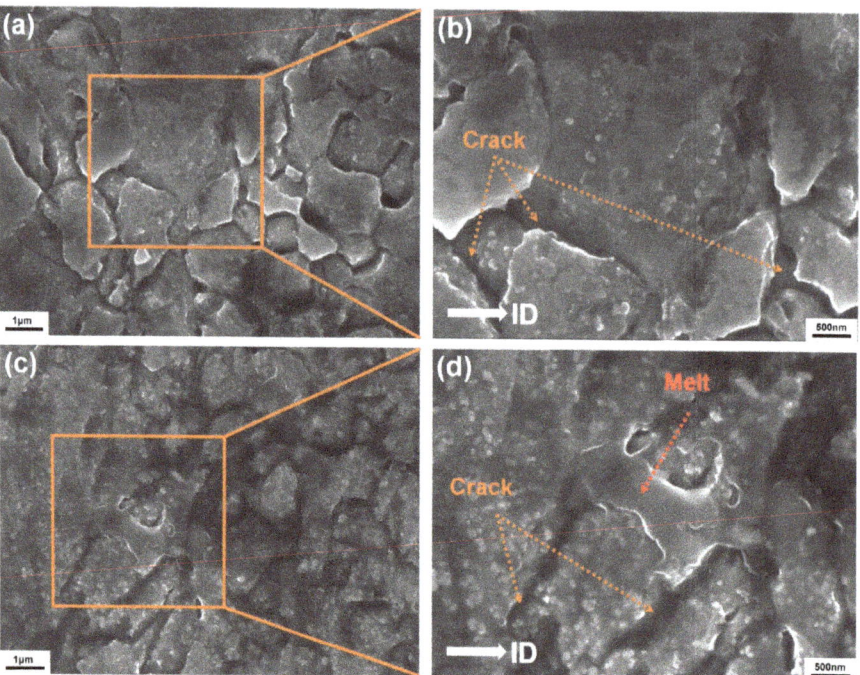

Figure 12. SEM morphology of adiabatic shear band: (**a,c**) SEM images inside the shear band; (**b,d**) are enlarged images of (**a,c**).

4. Conclusions

In this study, we conducted a comprehensive investigation of the dynamic impact behavior of AZ31 alloys with different rolling reductions and subjected them to different strain rates. Based on the above analysis and discussion of the microstructure and mechanical properties, the following conclusions can be drawn. With an escalation in strain rate, the formation of twins promoted DRX, and the microstructure became heterogeneous. These twins showed a remarkable effect regarding the modification of the orientation of neighboring grains. Importantly, at very high strain rates, the specimens experienced fractures, resulting in the formation of numerous adiabatic shear bands and deformation twins. In addition, a positive strain rate sensitivity was observed in the mechanical properties of the specimens prior to fracture, with the 20% rolling reduction alloy exhibiting the highest strength at a strain rate of 2000 s^{-1}. This increase in strength was attributed primarily to the combined influence of grain refinement, numerous dislocation formations, and profuse precipitates.

Author Contributions: Conceptualization, F.Y. and H.Y.; methodology, H.Y., W.Y., Y.X., B.J., K.S. and F.Y.; writing—original draft preparation, Y.L. (Yingjie Li), H.Y. and C.L.; writing—review and editing, H.Y., C.L., W.Y., Y.X., B.J., K.S. and F.Y.; validation, Y.L. (Yingjie Li), Y.L. (Yu Liu), C.L. and W.Y.; investigation, Y.L. (Yingjie Li), H.Y., C.L., Y.L. (Yu Liu) and W.Y.; data curation, Y.L. (Yingjie Li), W.Y., Y.L. (Yu Liu), Y.X., B.J. and K.S.; project administration, H.Y. All authors have read and agreed to the published version of the manuscript.

Funding: This work was supported by the Natural Science Foundation of Hebei province (no. E2022202158) and the foundation of strengthening program (2019-JCJQ-142-00).

Data Availability Statement: The raw/processed data required to reproduce these findings cannot be shared at this time, as the data are part of an ongoing study.

Conflicts of Interest: Author Yuling Xu was employed by the company Baosteel Metal Co., Ltd. The remaining authors declare that the research was conducted in the absence of any commercial or financial relationships that could be construed as a potential conflict of interest.

References

1. Zhao, Y.; Dong, G.; Zhao, B. Research progress of magnesium alloy application in aviation manufacturing. *Nonferrous Met. Eng.* **2015**, *5*, 23–27.
2. Yu, K.; Li, W.; Ma, Z. Research, development and application of wrought magnesium alloys. *Chin. J. Nonferrous Met.* **2003**, *13*, 277–288.
3. Yang, Y.; Li, J.; Song, H.; Liu, P. Application of Magnesium Alloys and Current Status of Their Forming Technology Research. *Hot Work. Technol.* **2013**, *42*, 24–27. (In Chinese)
4. Wang, J.; Ju, J.; Huang, Z.; Shu, Y. Research Progress on Preparation Technology of Magnesium Alloy Sheet. *Hot Work. Technol.* **2014**, *43*, 6–9+5.
5. Meng, R.; Zhang, D.; Yuan, H. Progress of Forming Technologies for Magnesium Alloy. *Hot Work. Technol.* **2008**, *37*, 89–92.
6. Hou, Z.; Jiang, B.; Wang, Y.; Song, J.; Xiao, L.; Pan, F. Development and Application of New Magnesium Alloy Materials and their New Preparation and Processing Technologies. *Aerosp. Shanghai* **2021**, *38*, 119–133.
7. Deng, H.; He, B. Research progress in fatigue properties of magnesium alloy welded joints. *Ordnance Mater. Sci. Eng.* **2016**, *39*, 125–129.
8. Chen, W.; Zhan, M.; Chen, W.; Zhang, D.; Li, Y. Present Status of Plastic Working for Wrought Magnesium Alloy and Its Future. *Spec. Cast. Non-Ferr. Alloys* **2007**, *27*, 40–43.
9. Bao, J.; Li, Q.A.; Chen, X.; Zhang, Q.; Chen, Z. Research Progress on Extruded Magnesium Alloys. *Mater. Rev.* **2022**, *36*, 20090073-12.
10. Elambharathi, B.; Kumar, S.D.; Dhanoop, V.U.; Dinakar, S.; Rajumar, S.; Sharma, S.; Kumar, V.; Li, C.; Eldin, E.M.T.; Wojciechowski, S. Novel insights on different treatment of magnesium alloys: A critical review. *Heliyon* **2022**, *8*, e11712. [CrossRef]
11. Che, B.; Lu, L.; Kang, W.; Luo, J.; Ma, M.; Liu, L. Hot deformation behavior and processing map of a new type Mg-6Zn-1Gd-1Er alloy. *J. Alloys Compd.* **2021**, *862*, 158700. [CrossRef]
12. Figueiredo, R.B.; Langdon, T.G. Deformation mechanisms in ultrafine-grained metals with an emphasis on the Hall–Petch relationship and strain rate sensitivity. *J. Mater. Res. Technol.-JmrT* **2021**, *14*, 137–159. [CrossRef]
13. Liu, Y.; Li, Y.; Zhu, Q.; Zhang, H.; Qi, X.; Wang, J.; Jin, P.; Zeng, X. Twin recrystallization mechanisms in a high strain rate compressed Mg-Zn alloy. *J. Magnes. Alloys* **2021**, *9*, 499–504. [CrossRef]
14. Long, J.; Xia, Q.; Xiao, G.; Qin, Y.; Yuan, S. Flow characterization of magnesium alloy ZK61 during hot deformation with improved constitutive equations and using activation energy maps. *Int. J. Mech. Sci.* **2021**, *191*, 106069. [CrossRef]
15. Malik, A.; Wang, Y.; Huanwu, C.; Nazeer, F.; Khan, M.A. Dynamic mechanical behavior of magnesium alloys: A review. *Int. J. Mater. Res.* **2019**, *110*, 1105–1115. [CrossRef]
16. Malik, A.; Nazeer, F.; Naqvi, S.Z.H.; Long, J.; Li, C.; Yang, Z.; Huang, Y. Microstructure feathers and ASB susceptibility under dynamic compression and its correlation with the ballistic impact of Mg alloys. *J. Mater. Res. Technol.-JmrT* **2022**, *16*, 801–813. [CrossRef]
17. Jin, Z.Z.; Cheng, X.M.; Zha, M.; Rong, J.; Zhang, H.; Wang, J.G.; Wang, C.; Li, Z.G.; Wang, H.Y. Effects of Mg17Al12 second phase particles on twinning-induced recrystallization behavior in Mg-Al-Zn alloys during gradient hot rolling. *J. Mater. Sci. Technol.* **2019**, *35*, 2017–2026. [CrossRef]
18. Zhang, L.; Townsend, D.; Petrinic, N.; Pellegrino, A. Measurement of Pure Shear Constitutive Relationship From Torsion Tests Under Quasi-Static, Medium, and High Strain Rate Conditions. *J. Appl. Mech.-Trans. ASME* **2021**, *88*, 121003. [CrossRef]
19. Zhou, S.; Deng, C.; Liu, S.; Liu, Y.; Zhu, J.; Yuan, X. Effect of strain rates on mechanical properties, microstructure and texture inside shear bands of pure magnesium. *Mater. Charact.* **2022**, *184*, 111686. [CrossRef]
20. Chen, Z.; Li, Q.; Chen, X.; Zhu, H. Research Status and Application of Zn-Containing Magnesium Alloys and Influence of LPSO on Alloy Properties. *J. Chin. Soc. Rare Earths* **2021**, *39*, 860–870.
21. Pan, H.; Ren, Y.; Fu, H.; Zhao, H.; Wang, L.; Meng, X.; Qin, G. Recent developments in rare-earth free wrought magnesium alloys having high strength: A review. *J. Alloys Compd.* **2016**, *663*, 321–331. [CrossRef]
22. Su, Z.; Huang, Y.; Liu, C.; Yang, X. Progress in RE-containing Cast Magnesium Alloys. *Spec. Cast. Nonferrous Alloys* **2015**, *35*, 1047–1051.
23. Cerreta, E.K.; Fensin, S.J.; Perez-Bergquist, S.J.; Trujillo, C.P.; Morrow, B.M.; Lopez, M.F.; Roach, C.J.; Mathaudhu, S.N.; Anghel, V.; Gray, G.T., III. The High-Strain-Rate Constitutive Behavior and Shear Response of Pure Magnesium and AZ31B Magnesium Alloy. *Metall. Mater. Trans. A-Phys. Metall. Mater. Sci.* **2021**, *52*, 3152–3170. [CrossRef]
24. Liu, F.; Liu, X.; Zhu, B.; Yang, H.; Xiao, G.; Hu, M. Influence of Microstructure and Mechanical Properties on Formability in High Strain Rate Rolled AZ31 Magnesium Alloy Sheets. *Met. Mater. Int.* **2022**, *28*, 1361–1371. [CrossRef]

25. Liu, X.; Wan, Q.; Yang, H.; Zhu, B.; Wu, Y.; Liu, W.; Tang, C. The Effect of Twins on Mechanical Properties and Microstructural Evolution in AZ31 Magnesium Alloy during High Speed Impact Loading. *J. Mater. Eng. Perform.* **2022**, *31*, 3208–3217. [CrossRef]
26. Yu, J.; Dong, F.; Xu, N.; Chen, Y.; Mao, P.; Liu, Z. Dynamic compressive properties and microstructural evolution of EW75 magnesium alloy at high temperatures and high strain rates. *Chin. J. Rare Met.* **2019**, *43*, 141–150.
27. Ji, Y.-F.; Duan, J.-R.; Yuan, H.; Li, H.-Y.; Sun, J.; Ma, L.-F. Effect of variable thickness cross rolling on edge crack and microstructure gradient of AZ31 magnesium alloy. *J. Cent. South Univ.* **2022**, *29*, 1124–1132. [CrossRef]
28. Yu, H.; Wang, D.; Liu, Y.; Liu, Y.; Huang, L.; Jiang, B.; Park, S.; Yu, W.; Yin, F. Recrystallization mechanisms and texture evolution of AZ31 alloy by gradient caliber rolling. *J. Mater. Res. Technol.* **2023**, *23*, 611–626. [CrossRef]
29. Guo, P.; Tang, Q.; Li, L.; Xie, C.; Liu, W.; Zhu, B.; Liu, X. The deformation mechanism and adiabatic shearing behavior of extruded Mg-8.0Al-0.1Mn alloy in different heat treated states under high-speed impact load. *J. Mater. Res. Technol.-JmrT* **2021**, *11*, 2195–2207. [CrossRef]
30. Li, Q. Mechanical properties and microscopic deformation mechanism of polycrystalline magnesium under high-strain-rate compressive loadings. *Mater. Sci. Eng. A-Struct. Mater. Prop. Microstruct. Process.* **2012**, *540*, 130–134. [CrossRef]
31. Zhu, B.; Liu, X.; Xie, C.; Liu, W.; Tang, C.; Lu, L. The flow behavior in as-extruded AZ31 magnesium alloy under impact loading. *J. Magnes. Alloys* **2018**, *6*, 180–188. [CrossRef]
32. Nazeer, F.; Naqvi, S.Z.H.; Kalam, A.; Al-Sehemi, A.G.; Alrobei, H. Texture dependencies on flow stress behavior of magnesium alloy under dynamic compressive loading. *Vacuum* **2021**, *191*, 110323. [CrossRef]
33. Tang, W.; Liu, S.; Liu, Z.; Kang, S.; Mao, P.; Zhou, L.; Wang, Z. Microstructure evolution and constitutive relation establishment of Mg–7Gd–5Y–1.2Nd–0.5Zr alloy under high strain rate after severe multi-directional deformation. *Mater. Sci. Eng. A-Struct. Mater. Prop. Microstruct. Process.* **2021**, *809*, 140994. [CrossRef]
34. Du, Y.; Du, W.; Zhang, D.; Ge, Y.; Jiang, B. Enhancing mechanical properties of an Mg–Zn–Ca alloy via extrusion. *Mater. Sci. Technol.* **2021**, *37*, 624–631. [CrossRef]
35. Wang, Q.; Zhai, H.; Xia, H.; Liu, L.; He, J.; Xia, D.; Yang, H.; Jiang, B. Relating Initial Texture to Deformation Behavior During Cold Rolling and Static Recrystallization Upon Subsequent Annealing of an Extruded WE43 Alloy. *Acta Metall. Sin.-Engl. Lett.* **2022**, *35*, 1793–1811. [CrossRef]
36. Xu, Y.; Yin, K.; Xia, L.; Chen, S.; Men, X.; Deng, T.; Wang, Y.; Zhang, S.-H. Study on High-Speed Tensile Mechanical Properties and Deformation Mechanism of 2195 Al-Li Alloy Sheet. *Rare Met. Mater. Eng.* **2022**, *51*, 1283–1292.
37. Liu, J.; Lu, L.; Zhong, Z. Deformation twins and annealing twins in high purity coarse-grained aluminum by equal channel angular pressing at high strain rate. *J. Mater. Eng.* **2021**, *49*, 89–94.
38. Deng, J.-F.; Tian, J.; Zhou, Y.; Chang, Y.; Liang, W.; Ma, J. Plastic deformation and fracture mechanisms of rolled Mg-8Gd-4Y-Zn and AZ31 magnesium alloys. *Mater. Des.* **2022**, *223*, 111179. [CrossRef]
39. Chen, Y.; Mao, P.; Wang, Z.; Cao, G. Tensile twin evolution of Mg–3Al–1Zn magnesium alloy during high-strain rate deformation. *Mater. Sci. Technol.* **2021**, *37*, 1452–1464. [CrossRef]
40. Gui, Y.; Cui, Y.; Bian, H.; Li, Q.; Ouyang, L.; Chiba, A. Role of slip and {10-12} twin on the crystal plasticity in Mg-RE alloy during deformation process at room temperature. *J. Mater. Sci. Technol.* **2021**, *80*, 279–296. [CrossRef]
41. Han, X.; Xiao, T.; Yu, Z. Microstructure, Texture Evolution, and Strain Hardening Behaviour of As-extruded Mg-Zn and Mg-Y Alloys under Compression. *J. Wuhan Univ. Technol.-Mater. Sci. Ed.* **2023**, *38*, 430–439. [CrossRef]
42. Deng, G.; Li, A.; Li, W.; Chang, G.; Liu, Y. Deformation Mechanism and Microstructural Evolution of a Mg–Y–Nd–Zr Alloy under High Strain Rate at Room Temperature. *J. Mater. Eng. Perform.* **2023**, *33*, 3101–3114. [CrossRef]
43. Dixit, N.; Xie, K.Y.; Hemker, K.J.; Ramesh, K.T. Microstructural evolution of pure magnesium under high strain rate loading. *Acta Mater.* **2015**, *87*, 56–67. [CrossRef]
44. Yang, Y.; He, J.; Huang, J.; Lian, X. Difference in adiabatic shear susceptibility between pure copper and Cu–30% Zn solid solution alloy at different strain rate. *J. Mater. Res.* **2023**, *38*, 1410–1419. [CrossRef]
45. Liu, X.Y.; Pan, Q.L.; He, Y.B.; Li, W.B.; Liang, W.J.; Yin, Z.M. Flow behavior and microstructural evolution of Al–Cu–Mg–Ag alloy during hot compression deformation. *Mater. Sci. Eng. A-Struct. Mater. Prop. Microstruct. Process.* **2009**, *500*, 150–154. [CrossRef]
46. Wei, Q.; Yuan, L.; Shan, D.; Guo, B. Study on the microstructure and mechanical properties of ZK60 magnesium alloy with submicron twins and precipitates obtained by room temperature multi-directional forging. *J. Mater. Sci.* **2023**, *58*, 13236–13250. [CrossRef]
47. Mottaghian, F.; Taheri, F. Strength and failure mechanism of single-lap magnesium-basalt fiber metal laminate adhesively bonded joints: Experimental and numerical assessments. *J. Compos. Mater.* **2022**, *56*, 1941–1955. [CrossRef] [PubMed]
48. Liu, Z.; Wu, F.; Feng, B.; Liu, L.; Dong, C.; Zhao, Y.; Song, B. Enhancing mechanical properties of friction stir welded AZ31 alloys by post-weld compression. *Sci. Technol. Weld. Join.* **2023**, *28*, 468–477. [CrossRef]
49. Ding, N.; Du, W.; Zhu, X.; Dou, L.; Wang, Y.; Li, X.; Liu, K.; Li, S. Roles of LPSO phases on dynamic recrystallization of high strain rate multi-directional free forged Mg-Gd-Er-Zn-Zr alloy and its strengthening mechanisms. *Mater. Sci. Eng. A-Struct. Mater. Prop. Microstruct. Process.* **2023**, *864*, 144590. [CrossRef]
50. Alaneme, K.K.; Okotete, E.A. Enhancing plastic deformability of Mg and its alloys—A review of traditional and nascent developments. *J. Magnes. Alloys* **2017**, *5*, 460–475. [CrossRef]
51. Zhu, S.Q.; Ringer, S.P. On the role of twinning and stacking faults on the crystal plasticity and grain refinement in magnesium alloys. *Acta Mater.* **2018**, *144*, 365–375. [CrossRef]

52. Tan, L.; Huang, X.; Wang, Y.; Sun, Q.; Zhang, Y.; Tu, J.; Zhou, Z. Activation Behavior of {10-12}-{10-12} Secondary Twins by Different Strain Variables and Different Loading Directions during Fatigue Deformation of AZ31 Magnesium Alloy. *Metals* **2022**, *12*, 1433. [CrossRef]
53. Doiphode, R.L.; Murty, S.V.S.N.; Prabhu, N.; Kashyap, B.P. Grain growth in calibre rolled Mg–3Al–1Zn alloy and its effect on hardness. *J. Magnes. Alloys* **2015**, *3*, 322–329. [CrossRef]
54. Pedersen, K.O.; Borvik, T.; Hopperstad, O.S. Fracture mechanisms of aluminium alloy AA7075-T651 under various loading conditions. *Mater. Des.* **2011**, *32*, 97–107. [CrossRef]
55. Zou, D.L.; Zhen, L.; Zhu, Y.; Xu, C.Y.; Shao, W.Z.; Pang, B.J. Deformed microstructure evolution in AM60B Mg alloy under hypervelocity impact at a velocity of 5 kms^{-1}. *Mater. Des.* **2010**, *31*, 3708–3715. [CrossRef]

Disclaimer/Publisher's Note: The statements, opinions and data contained in all publications are solely those of the individual author(s) and contributor(s) and not of MDPI and/or the editor(s). MDPI and/or the editor(s) disclaim responsibility for any injury to people or property resulting from any ideas, methods, instructions or products referred to in the content.

Article

Study on the Optimization of the Preparation Process of ZM5 Magnesium Alloy Micro-Arc Oxidation Hard Ceramic Coatings and Coatings Properties

Bingchun Jiang [1,2], Zejun Wen [2], Peiwen Wang [1], Xinting Huang [1], Xin Yang [1], Minghua Yuan [1] and Jianjun Xi [1,*]

1. School of Mechanical and Electrical Engineering, Guangdong University of Science and Technology, Dongguan 523083, China; jiangbingchun_2008@163.com (B.J.); 17825244707@163.com (P.W.); 13480058689@163.com (X.H.); 3230002360@student.must.edu.mo (X.Y.); 15879018326@163.com (M.Y.)
2. School of Mechanical and Electrical Engineering, Hunan University of Science and Technology, Xiangtan 411201, China; zjwen732@163.com
* Correspondence: xjj63@163.com

Abstract: Hard ceramic coatings were successfully prepared on the surface of ZM5 magnesium alloy by micro-arc oxidation (MAO) technology in silicate and aluminate electrolytes, respectively. The optimization of hard ceramic coatings prepared in these electrolyte systems was investigated through an orthogonal experimental design. The microstructure, elemental composition, phase composition, and tribological properties of the coatings were characterized by scanning electron microscopy (SEM), energy-dispersive X-ray spectroscopy (EDS), X-ray diffraction (XRD), and tribological testing equipment. The results show that the growth of the hard ceramic coatings is significantly influenced by the different electrolyte systems. Coatings prepared from both systems have shown good wear resistance, with the aluminate electrolyte system being superior to the silicate system in performance. The optimized formulation for the silicate electrolyte solution has been determined to be sodium silicate at 8 g/L, sodium dihydrogen phosphate at 0.2 g/L, sodium tetraborate at 2 g/L, and potassium hydroxide at 1 g/L. The optimized formulation for the aluminate electrolyte solution consists of sodium aluminate at 5 g/L, sodium fluoride at 3 g/L, sodium citrate at 3 g/L, and sodium hydroxide at 0.5 g/L.

Keywords: ZM5 magnesium alloy; micro-arc oxidation; process optimization; hard ceramic coating; ultra-anti-corrosion; tribology

1. Introduction

Magnesium (Mg) and its alloys have the advantages of low density, high specific strength, good electromagnetic shielding performance, etc., and have broad application prospects in various industries such as automotive, electronics, aviation, and aerospace [1–3]. Due to the poor corrosion resistance and low hardness of Mg and its alloys, its application in many fields is minimal. Common surface treatment technologies include micro-arc oxidation (MAO) [4,5], electroplating [6,7], thermal spraying [8,9], sol-gel [10], electrodeposition [11], and other technologies. Micro-arc oxidation (MAO) technology is one of the most effective surface treatment technologies to improve the surface hardness of light metal [12].

Micro-arc oxidation technology, also known as micro-plasma oxidation or anode spark deposition, is a surface treatment technology developed based on anodic oxidation technology. Its principle is characterized by the use of arc discharge to enhance and activate the reaction occurring on the anode to form a high-quality strengthened ceramic coating on the surface of the workpiece [13,14]. In this technology, the metal on the surface of the workpiece interacts with the electrolyte solution by applying a voltage to the workpiece through a unique micro-arc oxidation power supply to form a micro-arc discharge. Under the action of high temperature, electric field, and other factors, ceramic coating is formed on the metal surface to achieve the purpose of surface strengthening of the workpiece. In recent years, MAO technology has been successfully utilized to produce protective coatings on the surface of magnesium and its alloys, significantly enhancing the surface properties by increasing their resistance to wear, corrosion, and insulation. This advancement opens up new possibilities for the application of magnesium and its alloys. Furthermore, the MAO method can also be applied to process other metals and alloys such as aluminum and its alloys [15], titanium and its alloys [16], zinc alloys [17], etc., each with unique properties and potential uses. In micro-arc oxidation technology, the selection and optimization of electrolytic liquid systems are essential to achieve the best effect of the micro-arc oxidation technology on magnesium alloy. By adjusting parameters such as the composition [18], concentration [19], temperature [20], and pH value of the electrolyte [21], the formation and performance of the coating during the MAO process can be effectively controlled to meet the needs of different application scenarios. The properties of coatings prepared on magnesium and its alloys largely depend on the electrolyte's composition.

Du et al. [22] used micro-arc oxidation to prepare continuous and uniform dense coating under three electrolyte systems. They found that the phases, hardnesses, and friction factors of the three MAO coatings were significantly different, with the MAO coating layer prepared in the aluminate system having the highest roughness and hardness and the best wear resistance. Wang et al. [23] found that the thickness of the layer obtained in the sodium silicate electrolyte system was thicker than that of the sodium aluminate system and that the electrochemical corrosion resistance of the ceramic coatings obtained was significantly better than that of the sodium aluminates system through the cross-sectional appearance of the ceramic coatings. Muhaffel et al. [24] found that MAO coatings synthesized in aluminate electrolyte could not protect the AZ91 magnesium alloy from wear in corrosive media (0.9 wt.% NaCl solution) well compared to the dry sliding condition. Adding a certain amount of Na3PO4 to the acid electrolyte improved the corrosion resistance of the micro-arc oxidation coating of AZ91 magnesium alloy. Dong et al. [25] found that the thickness of the coatings obtained in the sodium aluminate electrolyte system was thicker than that in the sodium silicate electrolyte system and the electrochemical resistance of the coatings produced was significantly better than that in the sodium aluminate electrolyte system. Li et al. [26] showed an increase in abrasion resistance of the alloy by micro-arc oxidation with increased cathodic voltage in silicate electrolytes.

By adjusting the composition ratio of different electrolytes to change the ceramic coating's phase structure and thickness, the ceramic coating's corrosion resistance and wear resistance can be further affected. In this experiment, ZM5 was used as the research material to conduct an in-depth study of its electrochemical behavior during MAO. Through orthogonal experiments, the process parameters, electrolyte formula, and process flow were further optimized and the hard ceramic coatings preparation formula of silicate and aluminate electrolyte solution was explored. The oxide layer with high strength and high corrosion performance was prepared, which verified the feasibility and superiority of MAO technology of magnesium alloy and improved the engineering application prospect of ZM5 magnesium alloy. It provides a scientific basis for further optimization of process parameters.

2. Materials and Methods

2.1. Experimental Material and Coatings Preparation

The experimental material was ZM5 magnesium alloy, which was purchased commercially, provided by Shanghai Xuansheng Metal Product Co., Ltd. (Shanghai, China). and the mass fraction of each chemical component was Al 1.4%~2.0%, Zn 1.8%~2.8%, Mn 0.5%–0.68%, Si 5.0%~7.0%, Cu 0.03%, and the margin was Mg. The heat treatment state is quenching and artificial aging and its mechanical properties σ_b is 128 MPa, δ is 2.3%, and H is 89 HV.

The sample size is $\Phi 20$ mm $\times$ 5 mm, the through hole of $\Phi 3$ is processed above the sample, and the sample is preground by 400#, 800#, 1200#, 1500#, and 2000# water scrub to remove the oxide layer on the surface of the sample. The sample is fastened with $\Phi 3$ aluminum wire and immersed in electrolyte. The other end is connected to the positive electrode of the power supply and the negative electrode of the power supply is connected to stainless steel. The electrolytes are sodium silicate and sodium aluminate, two electrolytic liquid systems prepared with deionized water. The pH value is 8~12 at room temperature. A unique power supply for asymmetric bipolar pulse micro-arc oxidation was used in the experiment and the detailed parameters of the electrical parameters were set as shown in Table 1. The electrolyte temperature is controlled at 20~40 °C and the micro-arc oxidation time t is 90 min.

Table 1. Primary establishment of electrical parameters.

Argument	Value
power supply P (kW)	50
pulse waveform	rectangular wave
polarity	bipolar
output mode	constant current
forward current density i^+ (A/dm^2)	10
positive and negative pulse width ratio ε	1:1
pulse frequency f (Hz)	50
negative positive current density ratio J	1.3:1

2.2. Experimental Scheme Design

In this paper, the process optimization scheme of ZM5 magnesium alloy micro-arc oxidation hard ceramic coatings was designed for silicate and aluminate electrolytic liquid systems, respectively. An orthogonal experimental design was adopted and a four-factor and three-level orthogonal Table was selected. In the silicate system, the test factors were Na_2SiO_3 (2 g/L, 5 g/L and 8 g/L), Na_2HPO_4 (0.2 g/L, 0.4 g/L and 0.6 g/L), $Na_2B_4O_7$ (1 g/L, 2 g/L and 3 g/L), and KOH (0.5 g/L, 1.0 g/L and 1.5 g/L). In the aluminate system, the test factors were $NaAlO_2$ 2 g/L, 5 g/L and 8 g/L), NaF 2 g/L, 3 g/L and 4 g/L), $C_6H_5Na_3O_7$ 1 g/L, 2 g/L and 3 g/L), and KOH (0.5 g/L, 1.0 g/L and 1.5 g/L). All chemicals utilized in this study were supplied by Macklin (Shanghai, China). The orthogonal design test Table L9 (3^4) is selected and the L9 orthogonal test Tables of different electrolytic liquid systems are shown in Tables 2 and 3.

Table 2. Orthogonal experimental table $L_9(3^4)$ of silicate.

Number	Factor A: Na_2SiO_3 (g/L)	B: Na_2HPO_4 (g/L)	C: $Na_2B_4O_7$ (g/L)	D: KOH (g/L)
1	2	0.2	1	1
2	2	0.6	2	1.5
3	2	0.4	3	0.5
4	5	0.6	1	0.5
5	5	0.4	2	1
6	5	0.2	3	1.5
7	8	0.4	1	1.5
8	8	0.2	2	0.5
9	8	0.6	3	1

Table 3. Orthogonal experimental table $L_9(3^4)$ of aluminum chloride.

Number	Factor A: $NaAlO_2$ (g/L)	B: NaF (g/L)	C: $C_6H_5Na_3O_7$ (g/L)	D: KOH (g/L)
1	2	2	1	1
2	2	4	2	1.5
3	2	3	3	0.5
4	5	4	1	0.5
5	5	3	2	1
6	5	2	3	1.5
7	8	3	1	1.5
8	8	2	2	0.5
9	8	4	3	1

2.3. Performance Test and Tissue Observation

The thickness of hard ceramic coatings was measured by CTG-10 digital eddy current thickness gauge. The hardness of dense layer H of hard ceramic coatings was measured by an HVS-1000 digital microhardness tester, provided by Quan De Electronic Instrument Department (Xiameng, China). An HD-E808-60 salt spray testing machine, provided by Dongguan Haida Instrument Co. (Dongguan, China), was used for anti-corrosion rating Q and its standard was referred to the salt spray testing national standard GB/T2423.17 [27] rating judgment method. Precision balance is used to measure the ceramic coatings' wear capacity Δm and the Russian BPC-01 friction and wear testing machine, provided by Optimol Corporation (Ural, Russia), is used for hard ceramic coatings friction and wear test. The dual material is GCr15, the friction condition is dry friction, the relative sliding speed is 0.8m/s, and the radial load is 90N. DISCOVER X-ray diffractometer, S-4700 scanning electron microscope (SEM), provided by Bruker Physik-AG (Saarbrücken, Germany) and Hitachi Limited (Tokyo, Japan) respectively, and energy dispersive spectrometer (EDS) were used to test the phase and cross-section morphology of the ceramic coating.

3. Results and Analysis

3.1. Range Analysis and Optimization Results

Range analysis is used to optimize the level of each factor. The more significant the range is, the more substantial the influence of this factor on the test index [4]. The most critical difference indicates that among all the factors, this factor has a significant impact on the test index. This paper adopts the comprehensive balance method to optimize the analysis. Table 4 shows the orthogonal test results of the silicate system and aluminate system. Salt spray test rating Q and hard ceramic coatings thickness h were used as evaluation indexes. Tables 5 and 6 show the variance analysis results of salt corrosion resistance grade and ceramic coatings thickness index under silicate and aluminate electrolyte conditions, respectively. It can be judged that the optimal formula of silicate electrolyte is as follows: 8 g/L sodium silicate, 0.2 g/L disodium hydrogen phosphate, 2 g/L sodium tetraborate, and 1 g/L potassium hydroxide. The optimal formula of aluminate electrolyte is as follows: sodium aluminate 5 g/L, sodium fluoride 3 g/L, sodium citrate 3 g/L, and sodium hydroxide 0.5 g/L.

Table 4. Results of $L_9(3^4)$ orthogonal tests under different systems.

Number	Factor	Silicate System		Aluminate System	
		Q	h (μm)	Q	h (μm)
1		8	59	5	41
2		0	36	4	28
3		1	45	10	46
4		0	27	5	49
5		7	50	8	36
6		5	52	5	28
7		8	63	6	31
8		9	67	4	43
9		4	44	3	29

3.2. Effect of Two Electrolytic Liquid Systems on the Growth of Hard Ceramic Coatings

Figure 1 shows the concentration of sodium silicate and sodium aluminate, the fluctuation voltage (V_q) of micro-arc oxidation, and the hardness (H) of the coatings. V_q and H have the same trend with sodium silicate and sodium aluminate concentrations. As can be seen from Figure 1a, with the increase in sodium silicate and sodium aluminate concentrations, the corresponding fluctuating voltage decreases, which is caused by the rise in the conductivity of the electrolyte. When the concentration is 2 g/L, the fluctuation voltage of sodium aluminate electrolyte (475.3 V) is higher than that of sodium aluminate electrolyte (416.9V). With the increase in sodium silicate and sodium aluminate concentration, the arc voltage of the sodium aluminate electrolyte system drops faster than that of the sodium silicate electrolyte system. When the concentration of the two electrolytes is less than 5 g/L, the V_q of sodium aluminate electrolyte is greater than that of sodium silicate electrolyte. The situation is reversed when the concentration of the two electrolytes is greater than 5 g/L. It can be shown that the growth of hard ceramic coatings of magnesium alloys is significantly affected by the concentration of different electrolytic liquid systems. It can be seen from Figure 1b that the coating hardness of ZM5 magnesium alloy prepared under the two electrolytic liquid systems is higher. The hardness of hard coatings increases with the increase in sodium silicate and sodium aluminate. When the concentration is 2 g/L, the hardness of sodium aluminate hard coatings (461.7 HV) is lower than that of sodium aluminate hard coatings (577.2 HV). With the increased concentration of sodium silicate and sodium aluminate, the rise rate of sodium hard coatings H prepared by sodium aluminate electrolytic liquid is faster than that of sodium silicate electrolytic liquid.

Table 5. Data processing results of silicate corrosion resistance grade and ceramic film thickness indicators.

	Index	Na_2SiO_3 (g/L)	Na_2HPO_4 (g/L)	$Na_2B_4O_7$ (g/L)	KOH (g/L)
Q	Q_1	9	22	16	10
	Q_2	12	16	16	19
	Q_3	21	4	10	13
	q_1	3	7.33	5.3	3.3
	q_2	4	5.33	5.3	6.3
	q_3	7	1.33	3.3	4.33
	extreme/q'	3	6	2	3
	optimized result	8	0.2	1 and 2	1
h (μm)	h_1	46.7	59.3	49.67	46.33
	h_2	43	52.67	51	51
	h_3	58	35.67	47	50.33
	extreme/h'	15	23.63	4	4.67
	optimized result	8	0.2	2	1

Table 6. Data processing results of aluminate corrosion resistance grade and ceramic film thickness indicators.

	Index	$NaAlO_2$ (g/L)	Na_2HPO_4 (g/L)	$Na_2B_4O_7$ (g/L)	KOH (g/L)
Q	Q_1	19	14	16	19
	Q_2	18	24	16	16
	Q_3	13	12	18	15
	q_1	6.33	4.67	5.33	6.33
	q_2	6	8	5.33	5.33
	q_3	4.33	4	6	5
	extreme/q'	2	4	0.67	1.33
	optimized result	5	3	3	0.5
h (μm)	h_1	38.33	35.67	40.33	46
	h_2	37.67	37.33	35.67	35.33
	h_3	34.33	37.67	34.33	29
	extreme/h'	4	2	6	17
	optimized result	5	3	1	0.5

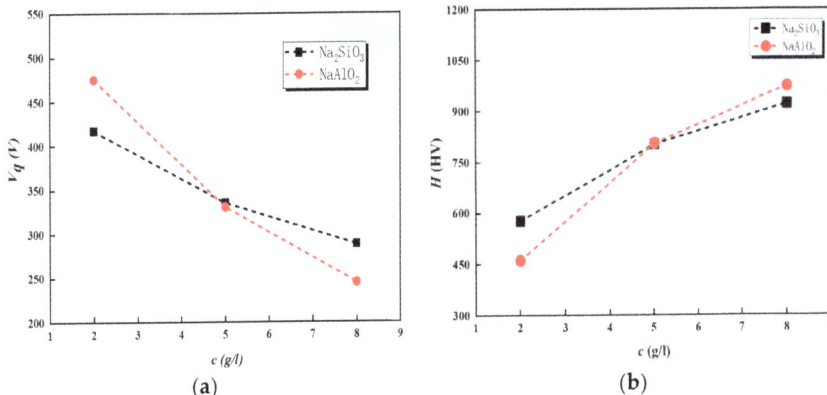

Figure 1. The relationship between Vq, H, and c under different electrolytic liquid systems (**a**) Vq-c; (**b**) H-c.

When the concentration is 5 g/L, the hardness of the coatings is close. When the concentration reaches 8 g/L, the hardness of the coatings exceeds 800 HV. It is concluded that the growth of hard ceramic coatings of magnesium alloys is also significantly affected by the concentration of different electrolytic liquid systems. Studies in the literature [4–6] show that electrolyte concentration positively correlates with conductivity. When the conductivity of the electrolyte is too large, the instantaneous energy of discharge will cause breakdown and damage in the micro-melt zone of the breakdown discharge, increasing defects in the ceramic coatings [28,29]. Therefore, in preparing hard ceramic coatings by micro-arc oxidation on the surface of ZM5 magnesium alloy, the sodium aluminate electrolytic liquid system is better than the sodium silicate electrolytic liquid system. When the concentration of sodium aluminate is 5~8 g/L, the hard ceramic coatings with high hardness and tiny pores can be prepared.

Figure 2 shows the coating thickness and hardness of the hard coatings under different electrolytic liquid systems. It can be seen from Figure 2a that the thickness of the micro-arc oxidation coatings gradually increases with the extension of oxidation time. At first, the growth rate of the ceramic coatings was speedy and then gradually slowed down until the thickness of the ceramic coatings no longer increased and the growth and dissolution rate of the ceramic coatings reached a relative dynamic balance. When t < 65 min, the long coatings velocity of ceramic coatings in the sodium silicate electrolytic liquid system is higher than that in the sodium aluminate electrolytic liquid system. When t > 65 min, the situation was reversed. This phenomenon may be related to the activity of silicon and aluminum elements and their atomic groups in the electrolytic liquid system [30,31]. It can be seen from Figure 2b that the hardness of the ceramic coating increases with the thickening of the dense layer (h_1) and the wear resistance of the coatings is correspondingly improved. It can also be seen from the figure that when h_1 > 35 μm, the hardness of ceramic coatings prepared by sodium aluminate electrolyte system is higher than that of sodium silicate electrolyte system. It is found in the test that when the forward current density i^+ < 5 A/dm^2, the energy of the breakdown moment is small, the dense layer phase transition energy is insufficient, and the hardness is low.

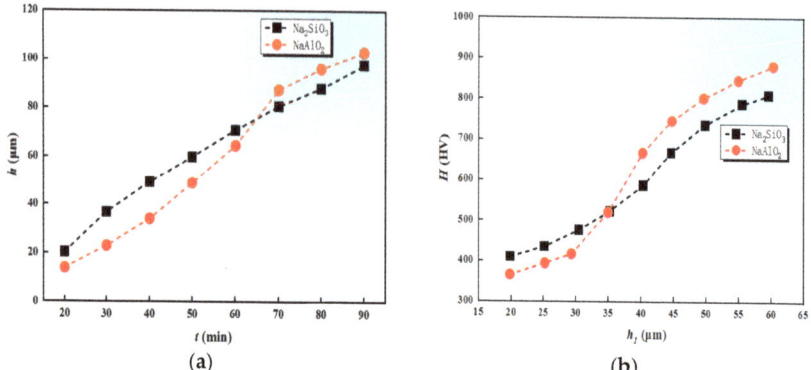

Figure 2. Coatings thickness and hardness of different electrolytic liquid systems: (**a**) h-t and (**b**) H-h_1.

3.3. Analysis of Hard Ceramic Coating Phase Structure and Profile Profile

Figure 3 shows the X-ray diffraction patterns of hard ceramic coatings prepared under different electrolytic liquid systems. As can be seen from Figure 3a, the compact layer of hard ceramic coatings prepared by the sodium silicate system is mainly composed of cube MgO and a small amount of spinel $MgAl_2O_4$, $MgSiO_3$, and Mg_2SiO_4. The main reaction equation is as follows:

$$Mg^{2+} + OH^- \rightarrow Mg(OH)_2 \rightarrow MgO + H_2O \tag{1}$$

$$Mg^{2+} + SiO_3^{2-} \rightarrow MgSiO_3 \tag{2}$$

$$2Mg^{2+} + SiO_3^{2-} + 2OH^- \rightarrow Mg_2SiO_4 + H_2O \tag{3}$$

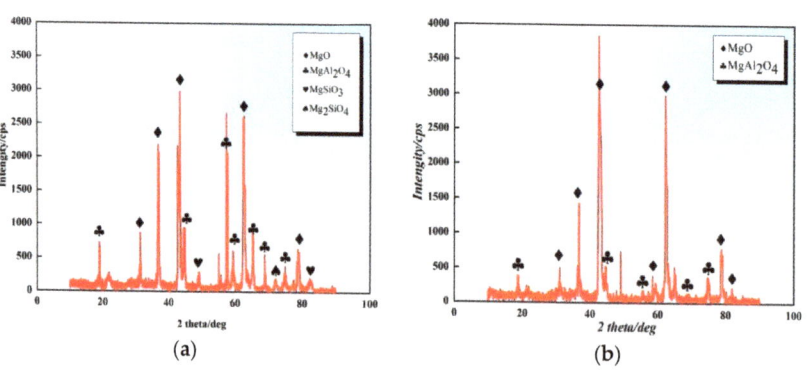

Figure 3. X-ray diffraction of hard ceramic coatings prepared with different electrolytic liquid systems: (**a**) silicate and (**b**) aluminates.

As shown in Figure 3b, the hard ceramic coatings prepared by the aluminate system mainly consist of cube MgO and spinel $MgAl_2O_4$. The primary reaction equation is as follows:

$$Mg^{2+} + OH^- \rightarrow Mg(OH)_2 \rightarrow MgO + H_2O \tag{4}$$

$$Mg^{2+} + 2AlO^- \rightarrow MgAl_2O_4 \tag{5}$$

The hard coatings prepared by the two electrolytic liquid systems contains a MgO phase with a larger crystal size due to sufficient crystal growth. Its corrosion resistance is greatly improved, while the hard ceramic coatings generated by sodium aluminate system micro-arc oxidation have better corrosion resistance, mainly because its crystal structure contains rich corrosion-resistant crystal—$MgAl_2O_4$. The properties of MAO ceramic coatings of magnesium alloys are significantly affected by the crystal phase. Specifically, the nature and dispersion pattern of the crystalline phase profoundly influence the microstructural organization and mechanical attributes of the coating, thereby exerting a significant impact on its corrosion resistance and mechanical robustness. In the MAO ceramic coating of magnesium alloy, the primary crystalline phase comprises oxides and the uniform dispersion of fine grains serves to enhance the density of the coating. Additionally, it enhanced the bonding strength between the oxide coatings and the substrate, ultimately increasing the corrosion resistance of the material.

Figure 4 shows the profile of the dense layer of hard ceramic coatings under different electrolytic liquid systems. It can be seen from the figure that the dense thickness of the hard coatings prepared by the silicate system and the aluminate system is ~69.6 μm and ~83.4 μm, accounting for 71.2% and 81.1% of the coatings' thickness, respectively. The density of the ceramic coatings prepared by the aluminate system is better than that of the silicate system and there is no obvious crack. In contrast, the ceramic coatings prepared by the silicate system have apparent cracks. This difference is mainly due to the high content of alumina in the aluminate system, which provides better thermal and chemical stability as well as better corrosion resistance. Moreover, during the preparation of the aluminate system, a layer of tightly bonded oxide layer was formed on the surface of aluminum oxide and magnesium alloy, which enhanced the adhesion and crack resistance of the coating.

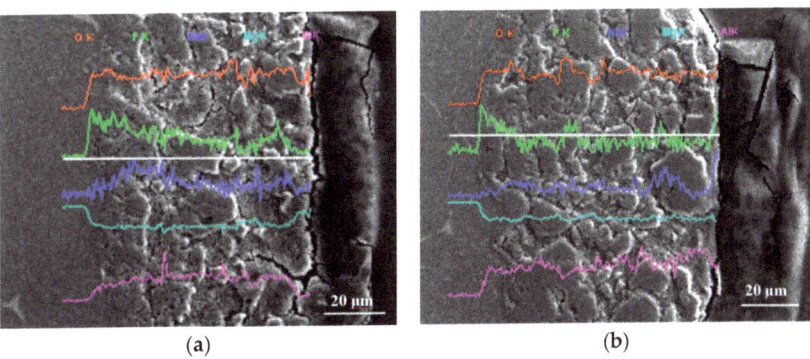

Figure 4. Cross-section topography and line analysis of hard ceramic coatings prepared with different electrolytic liquid systems: (**a**) silicate and (**b**) aluminates.

Figure 5 shows the corrosion resistance spectra of hard ceramic coatings prepared under different electrolytic liquid systems. It can be seen from Figure 5a that the elements contained in the hard ceramic coatings can be found in the electrolyte, among which the main elements are Mg, O, and Si, indicating that the main product of the corrosion-resistant hard ceramic coatings is MgO and $Mg_3[Si_4O_{10}](OH)_2$. As can be seen from Figure 5b, The crystal phase structure of the super anticorrosive coatings grown by MAO of aluminate electrolytic liquid system is mainly composed of MgO, Al_2O_3, and $MgAl_2O_4$ phases. Magnesium aluminum spinel ($MgAl_2O_4$) is a good performance of ceramic materials; its chemical properties are stable. When placed at room temperature, no acid or alkali reaction occurs and it has strong resistance to various melt erosion at high temperatures. Therefore, the MAO ceramic coating with Mg-Al spinel as the main phase has better corrosion resistance.

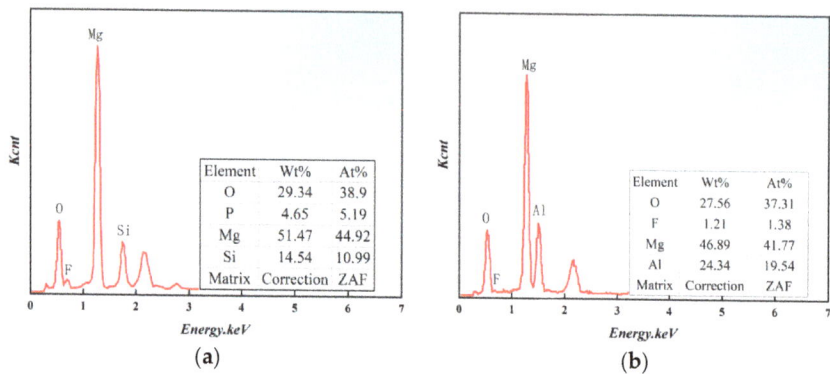

Figure 5. The energy spectrum of the sample of silicate system: (**a**) silicate and (**b**) aluminates.

3.4. Tribological Analysis of Hard Ceramic Coatings

The porous ceramic coatings generated by MAO can increase the contact area between the material and the surrounding environment, facilitate the diffusion of metal ions from the substrate, and further enhance corrosion resistance. This kind of ceramic coating has high hardness, good wear resistance, and corrosion resistance and can effectively resist friction and wear. Figures 6 and 7 show the relationship curves of wear quantity Δm—wear time t_m and μ—friction time t_c of ZM5 magnesium alloy micro-arc oxidation hard ceramic coatings, respectively. It can be seen from Figures 6 and 7 that Δm and μ decrease with the elongation of t_m and t_c because the generated ceramic coatings can be divided into a transition layer, dense layer, and loose layer from the inside out. The transition layer near the magnesium alloy matrix is metallurgically combined with the matrix. The dense layer mainly comprises MgO with good wear resistance and corrosion resistance, high hardness, and a small amount of $MgAl_2O_4$, Mg_2SiO_4, and $MgSiO_3$. The loose layer mainly comprises $MgAl_2O_4$, Mg_2SiO_4, and $MgSiO_3$ [32]. In addition, the microstructure and composition of ceramic coating grown on the MAO surface of magnesium alloy also have important effects on its wear resistance. The more uniform and dense the microstructure of the ceramic coating, the better the wear resistance. The tissue of the loose layer is loose so the wear amount and friction coefficient are significant. When the loose layer is removed, the wear amount begins to decrease gradually and the μ decreases accordingly and finally tends to be stable. The reason is that before 60 min, the loose layer of the ceramic coating gradually wears out. The decline rate of Δm and μ is fast, while after 60 min, the loose layer reaches the dense layer after wearing out and the density of the dense layer is improved and the hardness is increased, so the decline rate of Δm and μ becomes slow. The friction coefficient has a specific range for some commonly used wear-resistant materials, such as steel, plastics, rubber, etc. In these materials, the coefficient of friction below 0.3 is considered good tolerance. When the hard ceramic coating reaches 0.3, the coating time of the silicate system is about 20 min longer than that of the aluminate system. It can be inferred that the wear resistance of the loose layer of silicate coatings is more robust than that of the loose layer of aluminate coatings and the wear resistance of the dense layer of silicate coatings is more robust than that of the loose layer of aluminate coatings.

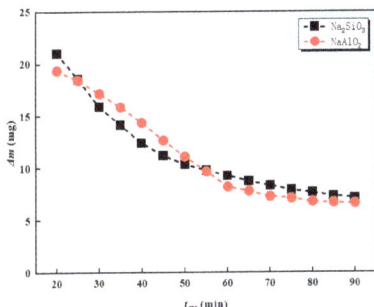

Figure 6. Relation between Δm and t_m.

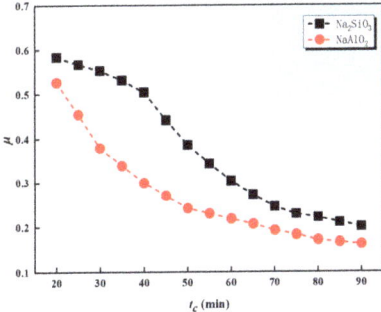

Figure 7. Relation between μ and t_c.

4. Conclusions

1. In preparing hard ceramic coatings by micro-arc oxidation on the surface of ZM5 magnesium alloy, the sodium aluminate electrolytic liquid system is better than the sodium silicate electrolytic liquid system. When the concentration of sodium aluminate is 5~8 g/L, the hard ceramic coatings with high hardness and tiny pores can be prepared;
2. The orthogonal experimental design method obtained the optimum process formula of micro-arc oxidation hard ceramic coatings under two systems. The optimal formula of silicate electrolyte is as follows: 8 g/L sodium silicate, 0.2 g/L disodium hydrogen phosphate, 2 g/L sodium tetraborate, and 1 g/L potassium hydroxide. The optimal formula of aluminate electrolyte is sodium aluminate 5 g/L, sodium fluoride 3 g/L, sodium citrate 3 g/L, and sodium hydroxide 0.5 g/L.
3. The coatings prepared by the two electrolytic liquid systems have good wear resistance. The wear resistance of the loose layer of silicate coatings is more robust than that of the loose layer of aluminate coatings and the wear resistance of the dense layer of silicate coatings is more robust than that of the dense layer of aluminate coatings.

Author Contributions: Conceptualization, B.J. and J.X.; methodology, B.J. and J.X.; validation, J.X. and Z.W.; formal analysis, B.J., P.W. and J.X.; investigation, P.W.; resources, B.J.; data curation, B.J., P.W., M.Y. and J.X.; writing—original draft preparation, B.J., P.W., X.H. and J.X.; writing—review and editing, B.J., Z.W., X.Y. and J.X.; visualization, X.H. and X.Y.; supervision, J.X.; project administration, B.J.; funding acquisition, B.J. and M.Y. All authors have read and agreed to the published version of the manuscript.

Funding: This research was funded by Guangdong Young Talent Innovation Programme (No 2022KQNCX116), the Dongguan Sci-Tech Commissioner Program (No. 20231800500602), the College Students Innovation and Entrepreneurship Training Program Project (202313719003), the Guangdong University of Science and Technology Innovative Research Team Project (GKY-2022CQTD-1), and the Project-Based Team of Teaching and Learning Through Teaching and Creating (GKJXXZ2023026).

Data Availability Statement: The original contributions presented in the study are included in the article, further inquiries can be directed to the corresponding author.

Conflicts of Interest: The authors declare no conflict of interest.

References

1. Rao, Y.Q.; Wang, Q.; Chen, J.X.; Ramachandran, C.S. Abrasion, sliding wear, corrosion, and cavitation erosion characteristics of a duplex coating formed on AZ31 Mg alloy by sequential application of cold spray and plasma electrolytic oxidation techniques. *Mater. Commun.* **2021**, *26*, 101978. [CrossRef]
2. Chen, Y.N.; Wu, L.; Yao, W.H.; Zhong, Z.Y.; Chen, Y.H.; Wu, J.H.; Pan, F.S. One-step in situ synthesis of graphene oxide Mg/Al-layered double hydroxide coating on a micro-arc oxidation coating for enhanced corrosion protection of magnesium alloys. *Surf. Coat. Technol.* **2021**, *413*, 127083. [CrossRef]
3. Ly, X.N.; Yang, S.; Nguyen, T.H. Effect of equal channel angular pressing as the pretreatment on microstructure and corrosion behavior of micro-arc oxidation (MAO) composite coating on biodegradable Mg-Zn-Ca alloy. *Surf. Coat. Technol.* **2020**, *395*, 125923. [CrossRef]
4. Xue, K.; Tan, P.H.; Zhao, Z.H.; Cui, L.Y.; Kannan, M.B.; Li, S.Q.; Liu, C.B.; Zou, Y.H.; Zhang, F.; Chen, Z.Y.; et al. In vitro degradation and multi-antibacterial mechanisms of β-cyclodextrin@curcumin embodied $Mg(OH)_2$/MAO coating on AZ31 magnesium alloy. *J. Mater. Sci. Technol.* **2023**, *132*, 179–192. [CrossRef]
5. Xia, Q.X.; Li, X.; Yao, Z.P.; Jiang, Z.H. Investigations on the thermal control properties and corrosion resistance of MAO coatings prepared on Mg-5Y-7Gd-1Nd-0.5Zr alloy. *Surf. Coat. Technol.* **2021**, *409*, 126874. [CrossRef]
6. Singh, C.; Tiwari, S.K.; Singh, R. Development of corrosion-resistant electroplating on AZ91 Mg Alloy by employing air and water-stable eutectic based ionic liquid bath. *Surf. Coat. Technol.* **2021**, *428*, 127881. [CrossRef]
7. Meng, X.; Wang, J.L.; Zhang, J.; Niu, B.L.; Gao, X.H.; Yan, H. Electroplated super-hydrophobic Zn-Fe coating for corrosion protection on magnesium alloy. *Trans. Nonferrous Met. Soc. China* **2022**, *32*, 3250–3258. [CrossRef]
8. Marzbanrad, B.; Razmpoosh, M.H.; Toyserkanic, E.; Jahed, H. Role of heat balance on the microstructure evolution of cold spray coated AZ31B with AA7075. *J. Magnes. Alloys* **2021**, *9*, 1458–1469. [CrossRef]
9. Palanisamy, K.; Gangolu, S.; Antony, J.M. Effects of HVOF spray parameters on porosity and hardness of 316L SS coated Mg AZ80 alloy. *Surf. Coat. Technol.* **2022**, *448*, 128898. [CrossRef]
10. Pereira, G.S.; Ramirez, O.M.P.; Avila, P.R.T.; Avila, J.A.; Pinto, H.C.; Miyazaki, M.H.; De Melo, H.G.; Bose Filho, W. Cerium conversion coating and sol-gel coating for corrosion protection of The WE43 Mg alloy. *Corrfosion Sci.* **2022**, *206*, 110527. [CrossRef]
11. Fattah-Alhosseini, A.; Chaharmahali, R.; Babaei, K. Effect of particles addition to solution of plasma electrolytic oxidation (PEO) on the properties of PEO coatings formed on magnesium and its alloys, a review. *J. Magnes. Alloys* **2020**, *8*, 799–818. [CrossRef]
12. Chaharmahali, R.; Fattah-Alhosseini, A.; Babaei, K. Surface characterization and corrosion behavior of calcium phosphate (Ca-P) base composite layer on Mg and its alloys using plasma electrolytic oxidation (PEO): A review. *J. Magnes. Alloys* **2021**, *9*, 21–40. [CrossRef]
13. Kaseema, M.; Fatimah, S.; Nashrah, N.; Ko, Y.G. Recent progress in surface modification of metals coated by plasma electrolytic oxidation: Principle, structure, and performance. *Prog. Mater. Sci.* **2021**, *117*, 100735. [CrossRef]
14. Lin, Z.S.; Wang, T.L.; Yu, X.M.; Sun, X.T.; Yang, H.Z. Functionalization treatment of micro-arc oxidation coatings on magnesium alloys: A review. *J. Alloys Compd.* **2021**, *879*, 160453. [CrossRef]
15. Chen, X.Y.; Zhang, Z.Q.; Duan, Y.W.; Wang, X.D. Growth mechanism of 2024 aluminum alloy micro-arc oxide layer in cobalt-containing electrolyte. *Surf. Coat. Technol.* **2023**, *462*, 129461. [CrossRef]
16. Chwartz, A.; Kossenko, A.; Zinigrad, M.; Danchuk, V.; Sobolev, A. Cleaning strategies of synthesized bioactive coatings by PEO on Ti-6Al-4V alloys of organic contaminations. *Materials* **2023**, *16*, 4624. [CrossRef] [PubMed]
17. Wang, S.K.; Meng, J.B.; Guan, Q.Y.; Dong, X.J.; Yu, H.Y.; Li, H.M.; Li, L. Preparation and performance analysis of CeO_2 particle-doped micro-arc oxidized composite film layer on the surface of copper-zinc alloy. *Surf. Technol.* **2023**, *52*, 152–161. [CrossRef]
18. Wu, J.H.; Wu, L.; Yao, W.H.; Chen, Y.N.; Chen, Y.H.; Yuan, Y.; Wang, J.F.; Atrens, A.; Pan, F.S. Effect of electrolyte systems on plasma electrolytic oxidation coatings characteristics on LPSO Mg-Gd-Y-Zn alloy. *Surf. Coat. Technol.* **2023**, *454*, 129192. [CrossRef]
19. Xie, P.; Blawert, C.; Serdechnova, M.; Konchakova, N.; Shulha, T.; Wu, T.; Zheludkevich, M.L. Effect of low concentration electrolytes on the formation and corrosion resistance of PEO coatings on AM50 magnesium alloy. *J. Magnes. Alloys* **2024**, *12*, 1386–1405. [CrossRef]
20. Wang, J.; Lu, H.L.; Sun, Z.B.; Xu, G.S.; Bai, Z.D.; Peng, Z.J. Effect of ultra accurate control of electrolyte temperature on the performance of micro arc oxidation ceramic coatings. *Ceram. Int.* **2023**, *49*, 33236–33246. [CrossRef]

21. Yao, W.H.; Zhan, G.X.; Chen, Y.H.; Qin, J.; Wu, L.; Chen, Y.N.; Wu, J.H.; Jiang, B.; Atrens, A.; Pan, F.S. Influence of pH on corrosion resistance of slippery liquid-infused porous surface on magnesium alloy. *Ftransactions Nonferrous Met. Soc. China* **2023**, *33*, 3309–3318. [CrossRef]
22. Du, C.Y.; Huang, S.T.; Yang, H.C.; Hu, Y.H.; Yu, X.L. Microstructure and properties of micro-arc oxidation films on SiC/Al matrix composites in different electrolyte systems. *J. Mater. Eng.* **2023**, *51*, 139–149.
23. Wang, Z.Y.; Ma, Y.; An, S.J.; Sun, L. Effect of electrolyte formulation on the corrosion resistance of micro-arc oxidation coating of pure magnesium. *Mater. Rep.* **2023**, *37*, 173–182.
24. Muhaffel, F.; Cimenoglu, H. Development of corrosion and wear resistant micro-arc oxidation coating on a magnesium alloy. *Surf. Coat. Technol.* **2019**, *357*, 822–832. [CrossRef]
25. Dong, H.R.; Li, Q.; Xie, D.B.; Jiang, W.G.; Ding, H.J.; Wang, S.; An, L.Y. Forming characteristics and mechanisms of micro-arc oxidation coatings on magnesium alloys based on different types of single electrolyte solutions. *Ceram. Int.* **2023**, *49*, 32271–32281. [CrossRef]
26. Li, X.J.; Zhangm, M.; Wen, S.; Mao, X.; Huo, W.G.; Guo, Y.Y. Microstructure and wear resistance of micro-arc oxidation ceramic coatings prepared on 2A50 aluminum alloys. *Surf. Coat. Technol.* **2020**, *394*, 125853. [CrossRef]
27. GB/T2423.17; Environmental testing for electric and electronic products: Part 2: Test Methods-Test Ka: Salt Mist. National Standards of People's Republic of China: Beijing, China, 2008.
28. Feng, Y.R.; Zhou, L.; Jia, H.Y.; Zhang, X.; Zhao, L.B.; Fang, D.Q. Research progress on micro-arc oxidation of medical magnesium alloy. *Surf. Technol.* **2023**, *52*, 11–24. [CrossRef]
29. Chen, H.T.; Li, Z.S.; He, Q.B.; Wu, H.L.; Ma, Y.Z.; Li, L. Study on artificial neural network used for optimization of micro-arc oxidation of magnesium alloys. *Rare Met. Mater. Eng.* **2012**, *41*, 23–27.
30. Guo, H.F.; An, M.Z. Growth of ceramic coatings on AZ91D magnesium alloys by micro-arc oxidation in aluminate–fluoride solutions and evaluation of corrosion resistance. *Appl. Surf. Sci.* **2005**, *246*, 229–238. [CrossRef]
31. Jia, Q.R.; Cui, H.W.; Zhang, T.T.; Cui, X.L.; Pan, Y.K.; Feng, R. General situation on research of micro-arc oxidation technology of magnesium alloys. *Mater. Prot.* **2018**, *51*, 108–113. [CrossRef]
32. Ma, N.; Huang, J.M.; Su, J.; Yin, L.Y. Effects of MgO nanoparticles on corrosion and wear behavior of micro-arc oxide coatings formed on AZ31B magnesium alloy. *Mater. Rep.* **2018**, *32*, 2768–2772. [CrossRef]

Disclaimer/Publisher's Note: The statements, opinions and data contained in all publications are solely those of the individual author(s) and contributor(s) and not of MDPI and/or the editor(s). MDPI and/or the editor(s) disclaim responsibility for any injury to people or property resulting from any ideas, methods, instructions or products referred to in the content.

Article

Improving the Corrosion Resistance of Micro-Arc Oxidization Film on AZ91D Mg Alloy through Silanization

Junchi Liu, Hang Yin, Zhengyi Xu *, Yawei Shao and Yanqiu Wang *

Key Laboratory of Superlight Materials and Surface Technology, Ministry of Education, Harbin Engineering University, Harbin 150001, China; liujunchi@hrbeu.edu.cn (J.L.); yinhang02@hrbeu.edu.cn (H.Y.); shaoyawei@hrbeu.edu.cn (Y.S.)
* Correspondence: xuzhengyi@hrbeu.edu.cn (Z.X.); qiuorwang@hrbeu.edu.cn (Y.W.)

Abstract: The presence of inherent micro-pores and micro-cracks in the micro-arc oxidation (MAO) film of Mg alloys is a key factor contributing to substrate corrosion. A composite film layer with high corrosion resistance was achieved through silanizing the micro-arc oxidation film. The corrosion performance of the MAO films treated with various silane coupling agents was assessed through morphological characterization and electrochemical tests. SEM graphs depicted that the silane film can effectively seal the defects existing in micro-arc oxidation film, and electrochemical tests indicated the significant corrosion resistance improvement of MAO film after silanization treatment.

Keywords: corrosion resistance; silanization; sealing; micro-arc oxidation film; AZ91D Mg alloy

Citation: Liu, J.; Yin, H.; Xu, Z.; Shao, Y.; Wang, Y. Improving the Corrosion Resistance of Micro-Arc Oxidization Film on AZ91D Mg Alloy through Silanization. *Metals* **2024**, *14*, 569. https://doi.org/10.3390/met14050569

Academic Editor: Petros E. Tsakiridis

Received: 10 April 2024
Revised: 5 May 2024
Accepted: 7 May 2024
Published: 12 May 2024

Copyright: © 2024 by the authors. Licensee MDPI, Basel, Switzerland. This article is an open access article distributed under the terms and conditions of the Creative Commons Attribution (CC BY) license (https://creativecommons.org/licenses/by/4.0/).

1. Introduction

Magnesium alloys are extensively employed in the aerospace sector, primarily for structural components, fuel tanks, and interior equipment of spacecraft, aiming to enhance performance and facilitate lightweight design [1,2]. Micro-arc oxidation (MAO) technology refers to the production of an oxidized film on metal surfaces through discharge plasma electrolysis treatment, which can significantly enhance the corrosion resistance and wear resistance of magnesium alloys [3]. Owing to its exceptional impact on surface modification and enhancement of metal material properties, MAO technology has found extensive applications in aerospace industries to protect the Mg alloy [4,5]. Nonetheless, the micro-arc oxide film may have some challenges, as the roughness and porosity of its surface could impact its corrosion and wear resistance in certain cases [6,7]. Thus, further surface modification of MAO films remains a key area of focus in current research [8].

Currently, to address the defects in MAO film, two primary methods have been developed to improve the integrity of the film: the one-step and two-step methods. The one-step method involves adding nanoparticles to the electrolyte during the formation of MAO film to increase its density and fill any micro-pores [9,10]. The two-step method involves building a composite film layer on the formed MAO film to repair any existing defects on the film surface through the external film layer [11]. The two-step method for micro-arc oxidizing is typically conducted through hydrothermal treatment [12,13], electroless plating [7,14], layered double hydroxide (LDH) [15], or other post-treatment. Although the external membrane layer of the composite film constructed through the two-step method is usually less abrasion-resistant and susceptible to mechanical damage compared to the traditional one-step method, it can provide better sealing and corrosion resistance, even after breakage [16,17]. Silanization treatment is a green surface treatment technology with high chemical resistance and thermal stability, which can effectively protect the surface of the substrate from environmental erosion and is among the methods employed. The formation of the silane film occurs through the dehydration of hydroxyl groups in M-OH on the metal surface and the reaction with silanol molecules or oligomers, resulting in the formation of M-O-Si bonds [18,19]. Additionally, the three-dimensional

(3D) network structure formed by cross-linked -Si-O-Si- bonds provides high resistance to corrosion [20,21]. A. Mandelli et al. [22] demonstrated the effectiveness of both Osi (Trimethyl (1-methylethenyloxy) silane) and BTSE (1,2-bis(trimethylsilyl)ethane) silane coupling agents in sealing the micro-pores of the MAO film layer and repairing any microscopic defects present in the AM60B MAO film.

This study organized a comparative investigation into the corrosion resistance of silane films on MAO surfaces and the improved corrosion resistance mechanism of MAO-silane composite film by utilizing various silane coupling agents as precursors. Several characterization tools, such as scanning electron microscopy (SEM), energy dispersive X-ray spectroscopy (EDS), and Fourier-transform infrared spectroscopy (FTIR), were utilized to analyze the morphology and composition of the MAO film surface. In addition, long-term EIS spectra are employed to analyze the prolonged corrosion behavior of composite coatings, thus discerning the failure modes of these composite coatings. Through analysis and discussion of the experimental results, the corrosion mechanism of the MAO-silane composite film was revealed, providing valuable guidance for practical engineering applications.

2. Materials and Methods

2.1. Preparation of MAO Film and Sealing Treatment

The chemical composition of the AZ91D magnesium alloy used in this study was analyzed using inductively coupled plasma atomic emission spectrometry (ICP-AES3600A, Nanjing, China). The analysis revealed the following composition (wt.%): Al 9.03, Zn 0.73, Mn 0.21, Si 0.002, Fe 0.001, Cu 0.001, Ni 0.003, with Mg as the balance element. The specimen, with dimensions of 40 mm × 35 mm × 5 mm, underwent gradual grinding to a 2000 # grit size. Subsequently, it was cleaned with deionized water and anhydrous ethanol and dried with cold air.

MAO treatment was performed on AZ91D magnesium alloy using a Doercoat II (Beijing, China) plasma electrolytic oxidation power supply, which includes an oxidation power supply, electrolyzer, electrode system, stirring system, and cooling system. An alkaline silicate system (5 g/L Na_2SiO_3, 4 g/L KO, and 4 g/L KF) was selected as the electrolyte, and the pulse oxidation current mode (with a current density of 8 A/dm^2, duty factor +50/−30, frequency 600 Hz) was used. The MAO film was formed by using the magnesium alloy workpiece as the anode and the graphite plate as the cathode, with the temperature controlled at 35 °C. The specimens with the MAO film were cleaned with deionized water and anhydrous ethanol, and then dried with cold air.

Three types of silane coupling agents—KH560 (γ-Glycidoxypropyltrimethoxysilane, contains an amino functional group and three ethoxy groups), BTSE (contains two silane groups and four sulfur atoms), and DTMS (dodecyltrimethoxysilane, contains two methyl groups and two ethoxy groups)—were chosen to prepare the silane composite film. The composite film was formed through a film-forming process (immersing the sample in silane hydrolysate at 50 °C for 20 min) and a curing process (at 120 °C in an oven for 1 h). The silane hydrolysate was prepared by mixing the silane coupling agent with ethanol and deionized water at a volumetric ratio of 1:1:4. The pH was adjusted to 4.0 with acetic acid, and the mixture was hydrolyzed for 8 h at 40 °C.

2.2. Electrochemical Tests

Electrochemical tests and neutral salt spray experiments were conducted to assess the corrosion resistance of the MAO and MAO-silane films. The electrochemical tests were performed in a 3.5 wt.% NaCl solution using the Zahner 6eX (Kronach, German) electrochemical workstation and a conventional three-electrode system. The saturated calomel electrode (SCE) served as the reference electrode, a platinum sheet as the counter electrode, and the exposed 1 cm^2 specimen as the working electrode. Potentiodynamic polarization tests were used to evaluate the corrosion resistance of the films. The potentiodynamic polarization tests were conducted in 3.5 wt.% NaCl solution within a potential range of −0.3 V vs. OCP to 1.6 V at a scan rate of 333 µV/s, with the tests being stopped when the anodic current reached 1 mA/cm^2. Electrochemical impedance spectroscopy (EIS) measurements were performed with a sinusoidal signal ranging from 10^{-2} to 10^5 Hz and an amplitude of 10 mV. The experimental data were analyzed using Zahner Analysis software. Prior to the tests, the system was allowed to stand for 10 min to stabilize the surface condition of the sample.

2.3. Salt Spray, Contact Angle Test, and Surface Characterization

The neutral salt spray tests were conducted in accordance with ASTM-B117 [23] using a salt spray tester (VSC/KWT1000, Ballingen, German) with a 5 wt.% NaCl solution. The specimen was coated with a molten mixture of paraffin and rosin to mitigate edge effects. The Barrington II contact angle tester was employed to assess the in situ conversion film of the magnesium alloy after silanization.

MAO films, the surface morphology after silanization, and the corresponding surface elemental distribution were analyzed using scanning transmission electron microscopy (FEI Titan Themis 200 TEM) and energy dispersive spectroscopy (Bruker Super-X EDS system, Billerica, MA, USA). Fourier transform infrared spectroscopy (FTIR) using a Nicolet 6700 instrument (Green Bay, WI, USA) was utilized to analyze the surface composition of the MAO-silane composite layer. As the sample of the composite membrane layer is solid, it was necessary to scrape off the membrane layer with a lancet, grind with a mortar, and mix it evenly with potassium bromide powder for pressing. The spectral range was set at 400 cm^{-1} to 4000 cm^{-1} with a resolution of 2 cm^{-1}, and the testing was repeated three times for each sample to ensure result accuracy.

3. Results

3.1. Contact Angle of MAO-Silanization Composite Film

The surface hydrophobicity results of the MAO film on AZ91D magnesium alloy after silanization treatment with KH560, BTSE, and DTMS are depicted in Figure 1. The contact angles (θ) of the MAO film treated with KH560, BTSE, and DTMS were 63.9°, 33.7°, and 115.6°, respectively. The high surface energy of the MAO film leads to rapid spreading of water droplets on its surface, rendering it challenging to measure the contact angle of the untreated film. Silanization of the MAO film can significantly lower its surface energy, enhance its hydrophobicity, and decrease the chances of contact between the corrosive medium and the film, thereby enhancing its corrosion resistance.

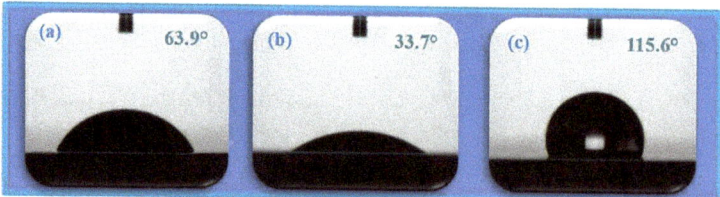

Figure 1. Contact angle of MAO film treated with (**a**) KH560, (**b**) BTSE, and (**c**) DTMS.

The choice of silane coupling agent significantly impacts the surface hydrophobicity of MAO-silane composite films. One end of the three silanes is Si-O-H, which binds to the hydroxyl group on the surface of the MAO film or other oligomers, forming a relatively dense silane film. The other end of the hydrolyzed product of BTSE is Si-O-H, while KH560 features an epoxide functional group that shares a similar polar bond with the O-H bond in a water molecule [8,24]. Consequently, the surface contact angle of BTSE and KH560 silane films is less than 90°. The hydrolyzed product of DTMS contains a hydrophobic long chain (dodecyl), leading to the formation of a predominantly hydrophobic silane film. As depicted in Figure 1, the surface hydrophobicity of the MAO film treated with the three silanes ranks in the order of DTMS > KH560 > BTSE.

3.2. Micro-Morphology

Figure 2 illustrates the surface morphology of the MAO film post-silanization with KH560, BTSE, and DTMS. Noticeable micro-pores and cracks were visible on the surface of the MAO film. Following silanization, some micro-cracks and small micro-pores on the MAO film surface were sealed, although the treatment had minimal impact on larger micro-pores. This could be attributed to the inability of the silane solution to penetrate the interior of the micro-pores, thus preventing the formation of a silane film within them. Consequently, the incomplete coverage of the silane film on the MAO film surface hinders the fundamental enhancement of the corrosion resistance of the MAO-silane composite film. Moreover, the acidic silane pre-hydrolysate may react with the exposed magnesium alloy through the defects in the MAO film, leading to the generation of corrosion products that impede the formation of a silane film within the micro-pores, compromising the membrane's integrity [25]. Nonetheless, the corrosion resistance of the silane film is not clearly discernible from Figure 2.

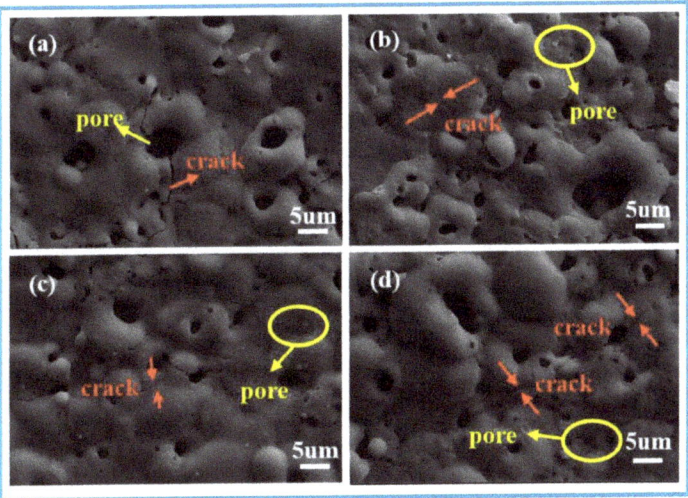

Figure 2. Surface morphology of MAO film modified by different silane coupling agents: (**a**) MAO; (**b**) MAO-KH560; (**c**) MAO-BTSE; (**d**) MAO-DTMS.

3.3. Electrochemical Measurements

Figure 3 depicts the potentiodynamic polarization curves of the MAO film both before and after silanization treatment with KH560, BTSE, and DTMS. Following the sealing treatment, a remarkable decrease in the corrosion current density of the MAO film and a positive shift in the corrosion potential were observed, indicating that all the silane coupling agents significantly contributed to the closure of structural defects in the MAO film, thereby enhancing its corrosion resistance. Furthermore, the MAO film treated with BTSE and DTMS exhibited a partial increase in the pitting potential of the film layer, which may be attributed to variations in the densification of the silane film. Additionally, the corrosion current density of MAO, MAO-KH560, MAO-BTSE, and MAO-DTMS film were 1.38×10^{-7}, 7.68×10^{-9}, 7.86×10^{-8}, and 1.05×10^{-8} A/cm^2, respectively, which demonstrated lower corrosion current density of the MAO-KH560 and MAO-DTMS composite films.

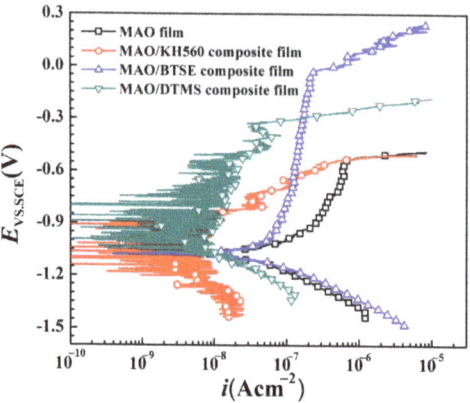

Figure 3. Potentiodynamic polarization curves of MAO film before and after treatment with different silane coupling agents.

The hydrolysis products of BTSE contain six Si-O-Hs, resulting in a relatively denser silane film compared to KH560 and DTMS. However, the highly hydrophilic surface of the BTSE silane film makes it more vulnerable to infiltration by corrosive media, which can penetrate the micro-defects of the MAO-silane composite film and reduce its long-term corrosion resistance. The slight repulsive effect of the hydrophobic long chains in DTMS on the hydroxyl groups results in the outer layer of the silane film being hydrophobic, leading to a more effective barrier against corrosive media.

Figure 4 illustrates the electrochemical impedance spectra of the AZ91D magnesium alloy MAO film before and after silanization with different silane coupling agents. The figure shows that the impedance modulus value, phase angle, and capacitance arc radius all increased after silanization, indicating that the silanization treatment can effectively enhance the integrity of the MAO film and improve its corrosion resistance. Therefore, the corrosion resistance of the MAO film is improved as a result. The results of the EIS test also laterally confirmed the results of the potentiodynamic polarization curves.

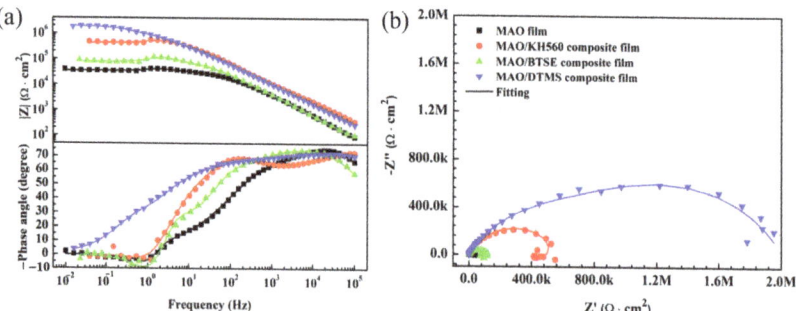

Figure 4. (a) Bode and (b) Nyquist diagrams of MAO film before and after silanization with different silane coupling agents.

3.4. Long-Term Immersion Experiments

As depicted in Figure 5, the corrosion resistance of the MAO film decreased with increasing immersion time. At 0.5 h immersion time, the impedance comprises a high-frequency capacitance arc, a medium-frequency capacitance arc, a low-frequency capacitance arc, and a low-frequency inductive reactance arc [26]. Initially, the corrosive medium reacts with the magnesium substrate at the film/substrate interface, and the impedance modulus decreases rapidly. With increasing immersion time, the impedance modulus continues to decrease. The process can be analyzed by data fitting using the equivalent circuit shown in Figure 6a, in which R_s represents solution resistance, and R_1 and R_2 represent the resistance of the loose and dense layers of the MAO film, respectively. CPE1 indicates the interaction between the outermost MAO film and the corrosive medium. The decrease in the peak phase angle predicts that the loose layer in the MAO film is infiltrated by the corrosive medium, leading to an increase in the conductivity of the MAO layer and a decrease in the resistance of the film layer. As the immersion time extended, the loose layer gradually failed. Similarly, CPE2 provides information on the dense MAO layer, and with increasing immersion time, this layer also gradually failed. CPE_{dl} indicates the corrosion occurring at the oxide film/substrate interface. The micro-defects on the surface of the MAO film provide channels for corrosive media to reach the MAO film/substrate interface. The low-frequency inductive reactance arc (L) indicates the occurrence of localized pitting corrosion on the AZ91D magnesium alloy substrate [27], and R_t means the charge transfer resistance.

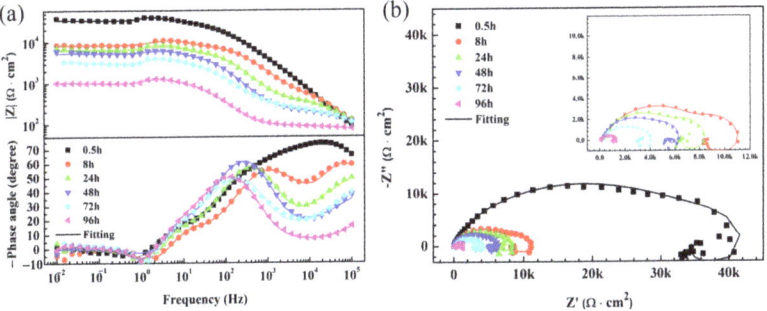

Figure 5. (**a**) Bode and (**b**) Nyquist diagrams for MAO film in 3.5 wt.% NaCl solution.

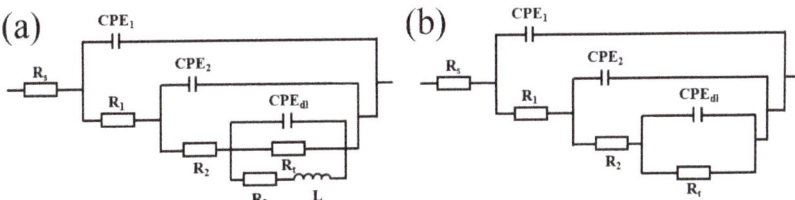

Figure 6. Two types of electrical equivalent circuit used for fitting the EIS spectra. (**a**) corrosion stage, (**b**) initial stage.

Figure 7 shows the EIS of MAO film treated with silane KH560, BTSE, and DTMS, respectively. In Figure 7a–c, an inductive reactance arc was not observed in the Nyquist spectra at an immersion time of 0.5 h, indicating that the corrosive medium did not reach the MAO/matrix interface. This proves that the silane treatment acts as a physical shielding effect on the corrosive medium, significantly improving the corrosion resistance of the MAO film. As the immersion process proceeded, a similar inductive reactance arc to that in Figure 5 appeared. When the immersion time reached 198 h, the loose layer of the MAO-silane film composite had been dissolved by the corrosive medium. Therefore, Figure 6b is used for fitting before 0.5 h, and Figure 6a is utilized when an inductive reactance arc appears [28].

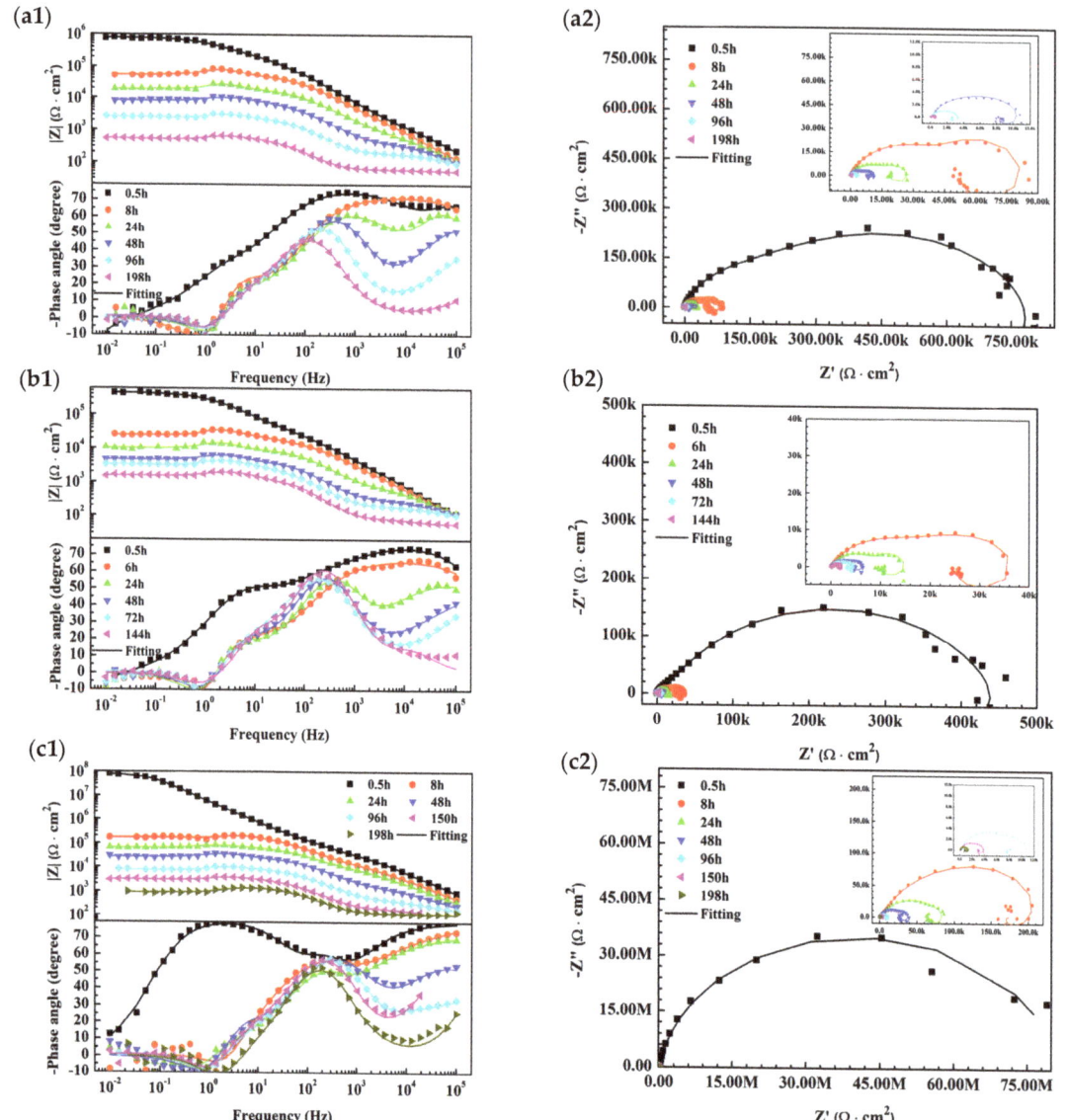

Figure 7. (**1**) Bode and (**2**) Nyquist diagram of AZ91D Mg alloy MAO film immersing in 3.5 wt.% NaCl solution for different time after silanization with different silane coupling agents: (**a**) KH560; (**b**) BTSE; (**c**) DTMS.

The low-frequency impedance modulus values ($|Z|_{0.01Hz}$) can be used to evaluate the protective coating failure process [28]. Figure 8 shows $|Z|_{0.01Hz}$ before and after the silanization treatment of the MAO film. The modulus satisfies MAO-DTMS > MAO-KH560 > MAO-BTSE > MAO film, which is consistent with the results of polarization tests. During the immersion process, the corrosive medium passes through the silane film and reacts with the magnesium alloy substrate, resulting in a gradual decrease in $|Z|_{0.01Hz}$ of the composite layer. Additionally, during the early stage (stage I) of immersion (0.5–8 h), the $|Z|_{0.01Hz}$ of the MAO-silane composite layer decreased rapidly from 10^6–10^7 Ω·cm^2 to 10^4–10^5 Ω·cm^2. From 8–96 h of immersion (stage II), the decrease in the $|Z|_{0.01Hz}$ slowed down with prolonged immersion time due to the gradual accumulation of corrosion products in the micro-pores. This accumulation hindered the mass transfer process of the corrosive medium, thus delaying the corrosion reaction. The $|Z|_{0.01Hz}$ was further reduced in stage III, which may be the reason for the gradual complete failure of the silane film.

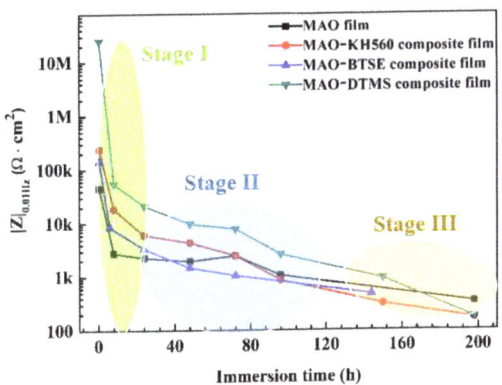

Figure 8. Time dependence of impedance modulus at 0.01 Hz.

The fitting results of R_1 and R_2 with immersion time are shown in Figure 9. The trend of R_1 and R_2 with immersion time is similar. During the initial stage of immersion (0.5 h), the R_1 and R_2 values for the MAO-silane film are 1–2 orders of magnitude higher than those for the MAO film, indicating a significant enhancement in the physical shielding capability of the MAO film due to silanization treatment. In the middle phase of immersion (8–48 h), R_1 and R_2 gradually decrease with immersion time, as the silane film hinders the mass transfer process of the corrosive medium. As the immersion time reaches 48–72 h, the resistance values of R_1 and R_2 decrease significantly once more, likely because the silane film is hydrolyzed by the corrosive medium, leading to a reduction in the density of the silane film.

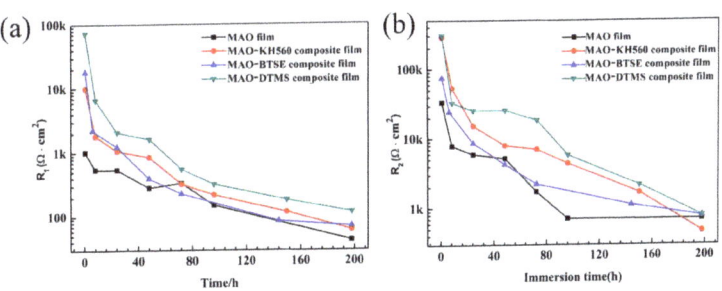

Figure 9. Time dependence of (**a**) R_1 and (**b**) R_2 of MAO film before and after silanization.

A higher resistance value of the charge transfer resistance (R_t) indicates greater difficulty in charge transfer at the interface between the magnesium alloy substrate and the corrosive medium [29]. In Figure 10, during the initial stage of immersion (0.5 h), the R_t value for MAO-DTMS was 10^8 $\Omega \cdot cm^2$, while the R_t values for MAO-KH560 and MAO-BTSE composite layers ranged from 10^5 to 10^6 $\Omega \cdot cm^2$, significantly higher than that of the MAO film. From 0.5 to 8 h, R_t decreased rapidly, indicating the initial degradation of the silane film layer. As the immersion time progresses, the decrease in R_t may be attributed to the accumulation of corrosion products at the interface, leading to the hindrance of charge transfer. However, as the immersion time further extends, the corrosion channel widens, facilitating charge transfer.

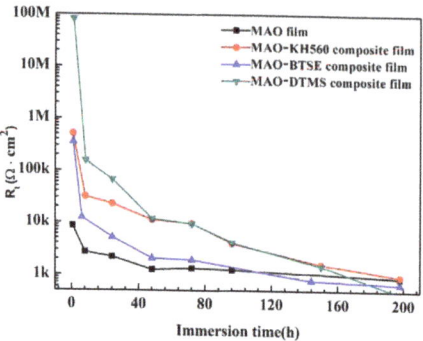

Figure 10. Time dependence of Rt of MAO film before and after silanization.

The EIS analysis showed that silanization treatment effectively enhances the corrosion resistance of the MAO film, but the protective effect of the silane film diminishes gradually with prolonged immersion time.

3.5. Long-Term Salt Spray Experiments

Figure 11 illustrates the macroscopic morphology of the MAO specimen surface under neutral salt spray conditions before and after silanization treatment. The red circles in Figure 11 represent the locations at which the visible corrosion features are first observed, caused by the accumulation of corrosion products on the surface of the MAO film or small pits, which gradually change into large areas of corrosion products in the course of time. The untreated MAO film exhibited its initial corrosion point in the neutral salt spray environment at 24 h. As the salt spray duration increased, both the number of corrosion points and the corrosion area expanded. By 96 h, the MAO-KH560 and MAO-BTSE composite films displayed their first corrosion point, with the number of corrosion points and corrosion area remaining consistent with the duration of the salt spray exposure. The MAO film treated with DTMS silanization did not exhibit corrosion in the salt spray environment until 552 h, demonstrating superior resistance to salt-spray-induced corrosion.

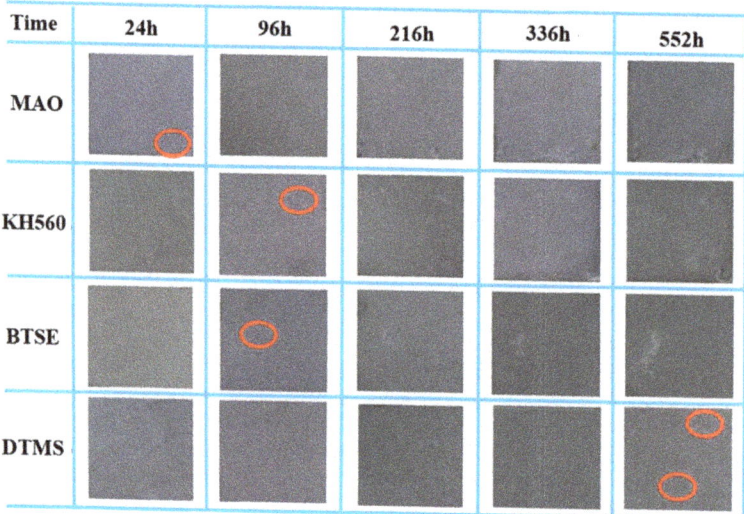

Figure 11. The macroscopic morphology of MAO film tested by neutral salt spray at different time before and after treatment with different silane coupling agents.

Figure 12 illustrates the micro-surface morphology of the MAO-silane composite film after the neutral salt spray test. Following the salt spray test, the micro-arc oxidation film underwent significant changes in surface morphology, resulting in the formation of a layer of loose and porous flaky corrosion products on the original surface. The surface of the MAO-DTMS composite film layer did not exhibit noticeable changes, possibly due to the effective surface hydrophobicity of the DTMS silane film, which inhibited the occurrence of corrosion reaction. Volcano-like protrusions of corrosion products were observed on the surface of the MAO-KH560 and MAO-BTSE composite film layers, indicating that the surface of the KH560 or BTSE silane membranes was incomplete, leading to the occurrence of corrosion.

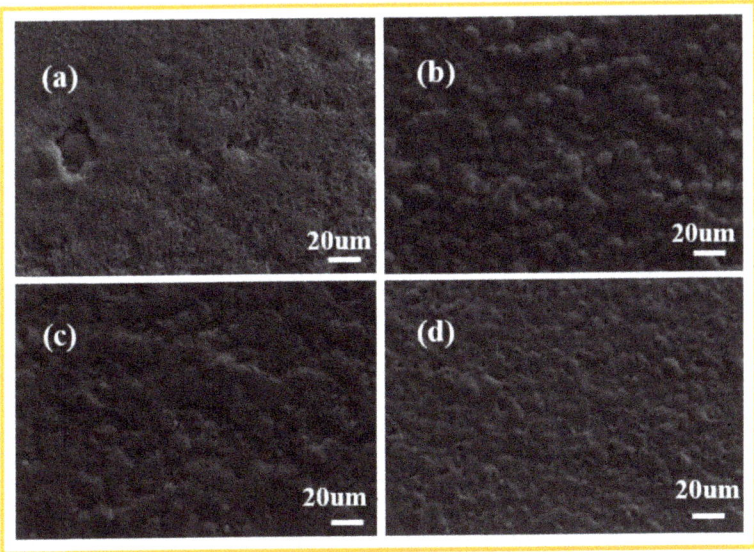

Figure 12. The microscopic surface morphology of MAO film after salt spray test before and after treatment with different silane coupling agents: (**a**) MAO coating; (**b**) KH560; (**c**) BTSE; (**d**) DTMS.

4. Discussion

Figure 2 displays the surface of the AZ91D magnesium alloy MAO film, revealing distinct structural defects such as micro-holes and micro-cracks. In the micro-arc oxidation process, the application of high voltage causes breakdown in the thinnest region of the oxide film, leading to the ejection of high-temperature molten oxide along the discharge channel under the influence of the electric and pressure fields. Quenching occurs when the molten oxide encounters the electrolyte, resulting in the formation of a "crater"-like microscopic morphology and generating stresses that lead to the formation of micro-cracks.

Figure 13 depicts the cross-sectional morphology and elemental surface distribution of the MAO film before and after the silanization treatment. In Figure 13a, the MAO film is divided into an outer porous layer and an inner dense layer. The outer layer is thicker and more porous, whereas the inner layer is thinner but denser, exhibiting fewer defects. Furthermore, the elemental distribution diagram of the MAO film cross-section indicates that the film is predominantly composed of Mg, O, and Si. In Figure 13b, a "Si"-rich film layer is observed on the surface of the MAO film, indicating that the silanization treatment can effectively create a protective silane film on the MAO film. Moreover, significant enrichment at the micro-cracks was observed, indicating that silane can be deposited within the micro-cracks of the MAO film, enhancing the integrity of the MAO membrane [30].

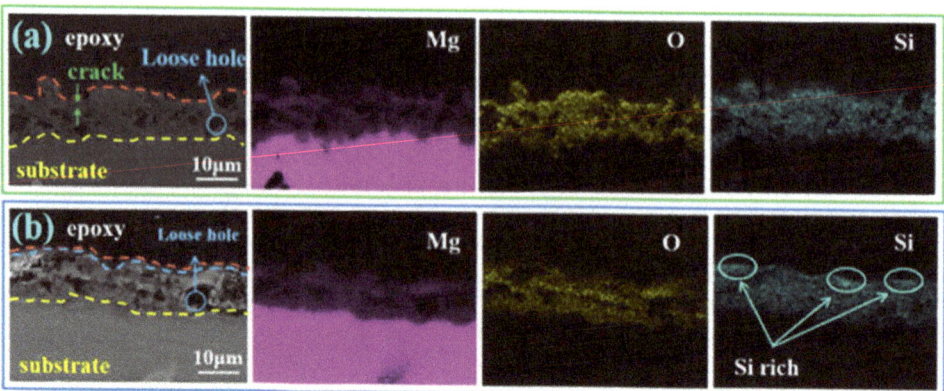

Figure 13. Micromorphology of the cross-section of MAO film (**a**) before and (**b**) after DTMS silanization.

Figure 14 displays the FT-IR spectra of the magnesium alloy MAO film before and after the silanization treatment with DTMS. The peak at 3653 cm^{-1} in the MAO film corresponds to the metal hydroxide O-H stretching vibration peak [31], while the peak at 1008 cm^{-1} indicates the presence of Mg_2SiO_4 in the MAO film [32]. Following silanization, peaks were observed at 2932 cm^{-1}, 2858 cm^{-1}, 1101 cm^{-1}, and 903 cm^{-1}, corresponding to the antisymmetric CH_2, symmetric CH_2, Si-O-S, and Si-OH stretching vibrations in the silane film [33], respectively. Notably, the peaks at 3653 cm^{-1} and 1008 cm^{-1} persist after the silanization of the MAO film, possibly indicating the destruction of the MAO film during sample preparation or the incomplete sealing of the MAO film by the silanization process.

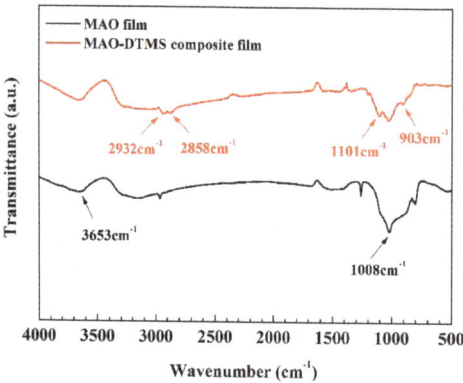

Figure 14. FTIR spectra of MAO film before and after DTMS silanization.

The corrosion mechanism of the MAO-silane composite film in a 3.5 wt.% NaCl solution is illustrated in Figure 15 based on the preceding analysis. This mechanism can be categorized into three stages based on Figure 7. In Stage I, the corrosive medium is obstructed by the silane film covering the MAO surface, providing effective corrosion resistance. Despite some corrosive medium penetrating the film through defects, the corrosion rate remains relatively low, as depicted in Figure 15a. As the immersion time increases, corrosion at the interface between the magnesium alloy substrate and the MAO intensifies, preventing timely outward diffusion of corrosion products. This results in the gradual accumulation of corrosion products in the pores, consequently decelerating the corrosion rate, as shown in Figure 15b. In Stage III, the surface silane film undergoes gradual hydrolysis and dissolution, allowing corrosion products to diffuse outward through the exposed defects. This re-exposes the magnesium alloy substrate at the defects to the corrosive medium, accelerating the corrosion process, as illustrated in Figure 15c.

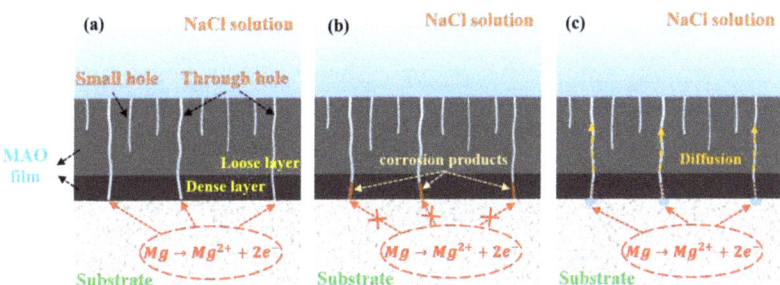

Figure 15. Schematic diagram of corrosion process of MAO-silane composite film. (**a**) initial stage, (**b**) corrosion product accumulation, (**c**) diffusion stage.

5. Conclusions

The silanization treatment serves as a pivotal method in effectively sealing minute defects within the magnesium alloy MAO film, the hydrolysis product of the silane coupling agent plays a crucial role in determining the hydrophilicity and density of the resultant silane film. The degree of crosslinking in the silane film can enhance its sealing effect on defects within the MAO film, while hydrophobicity can resist the intrusion of corrosive media at minor defect sites, thereby substantially augmenting its long-term corrosion resistance. Consequently, the choice of silane coupling agent profoundly impacts the defect-sealing efficacy of the MAO film and the corrosion resistance of the composite layer. Particularly noteworthy, KH560 and DTMS silanization treatments manifest a notably enhanced improvement in the corrosion resistance of the MAO film. This enhancement is evidenced by a nearly two orders of magnitude reduction in the corrosion current density, accompanied by a significant increase in the value of $|Z|_{0.01Hz}$.

Author Contributions: Conceptualization, J.L. and H.Y.; methodology, Y.W.; software, Y.S.; validation, Z.X., Y.S. and Y.W.; formal analysis, J.L. and H.Y.; investigation, Y.W.; resources, Y.S. and Y.W.; data curation, Z.X.; writing—original draft preparation, J.L. and H.Y.; writing—review and editing, Z.X.; visualization, Z.X.; supervision, Y.W.; project administration, Y.S. and Y.W.; funding acquisition, Y.W. All authors have read and agreed to the published version of the manuscript.

Funding: This research was funded by National Natural Science Foundation of China (grant numbers: 51971073, 52001155); Postdoctoral funding project of Heilongjiang Province (grant number: LBH-Z23122), and the Natural Science Foundation of Jiangxi Province (grant number: 20212BAB214038).

Data Availability Statement: The raw data supporting the conclusions of this article will be made available by the authors on request.

Conflicts of Interest: The authors declare no conflicts of interest.

References

1. Tang, X.; Zhang, Q.; Mei, D.; Liu, M.; Wang, L.; Zhu, S.; Guan, S. The uniform corrosion of biodegradable Mg alloy induced by protein addition in Hanks' balanced salt solution. *Colloids Surf. A* **2024**, *690*, 133824. [CrossRef]
2. Song, Y.; Dai, J.; Sun, S. A comparative study on the corrosion behavior of AZ80 and EW75 Mg alloys in industrial atmospheric environment. *Mater. Today Commun.* **2024**, *38*, 108263. [CrossRef]
3. Wu, M.; Jiang, F. Preparation, interface properties and corrosion behavior of nano-modified MAO ceramic film on 5B70 Al alloy. *J. Alloys Compd.* **2023**, *967*, 171829. [CrossRef]
4. He, R.Y.; Wang, B.Y.; Xiang, J.H.; Pan, T.J. Effect of copper additive on microstructure and anti-corrosion performance of black MAO films grown on AZ91 alloy and coloration mechanism. *J. Alloys Compd.* **2021**, *889*, 161501. [CrossRef]

5. Gong, Y.; Geng, J.; Huang, J.; Chen, Z.; Wang, M.; Chen, D.; Wang, H. Self-healing performance and corrosion resistance of novel CeO2-sealed MAO film on aluminum alloy. *Surf. Coat. Tech.* **2021**, *417*, 127208. [CrossRef]
6. Lee, H.-B.; Sheu, H.-H.; Jian, J.-S.; Hsiao, R.-C. Study on the Characteristics of MAO/Polymer/Ni Three-Layer Composite Film formed on AZ31 Magnesium Alloy. *Int. J. Electrochem. Sci.* **2021**, *16*, 211246. [CrossRef]
7. Lee, C.-Y.; Lee, J.-L.; Jian, S.-Y.; Chen, C.-A.; Aktuğ, S.L.; Ger, M.-D. The effect of fluoride on the formation of an electroless Ni–P plating film on MAO-coated AZ31B magnesium alloy. *J. Mater. Res. Technol.* **2022**, *19*, 542–556. [CrossRef]
8. Zhu, J.; Jia, H.; Liao, K.; Li, X. Improvement on corrosion resistance of micro-arc oxidized AZ91D magnesium alloy by a pore-sealing coating. *J. Alloys Compd.* **2021**, *889*, 161460. [CrossRef]
9. Wang, S.-Y.; Si, N.-C.; Xia, Y.-P.; Li, L. Influence of nano-SiC on microstructure and property of MAO coating formed on AZ91D magnesium alloy. *Trans. Nonferrous Met. Soc. China* **2015**, *25*, 1926–1934. [CrossRef]
10. Cui, L.; Wang, Y.; Hu, L.; Gao, L.; Du, B.; Wei, Q. Mechanism of Pb (II) and methylene blue adsorption onto magnetic carbonate hydroxyapatite/graphene oxide. *RSC Adv.* **2015**, *5*, 9759–9770. [CrossRef]
11. Liu, R.; Xu, D.; Liu, Y.; Wu, L.; Yong, Q.; Xie, Z.-H. Enhanced corrosion protection for MAO coating on magnesium alloy by the synergism of LDH doping with deposition of 8HQ inhibitor film. *Ceram. Int.* **2023**, *49*, 30039–30048. [CrossRef]
12. Zhang, Y.; Luo, S.; Wang, Q.; Ramachandran, C.S. Effect of hydrothermal treatment on the surface characteristics and bioactivity of HAP based MAO coating on Ti-6Al-4V alloy. *Surf. Coat. Tech.* **2023**, *464*, 129566. [CrossRef]
13. Ma, F.C.; Liu, P.; Chen, Y.; Li, W.; Liu, X.K.; Chen, X.H.; He, D.H. Various Morphologies Hydroxyapatite Crystals on Ti MAO Film Prepared by Hydrothermal Treatment. *Phys. Procedia* **2013**, *50*, 442–448. [CrossRef]
14. Chen, C.-A.; Jian, S.-Y.; Lu, C.-H.; Lee, C.-Y.; Aktuğ, S.L.; Ger, M.-D. Evaluation of microstructural effects on corrosion behavior of AZ31B magnesium alloy with a MAO coating and electroless Ni-P plating. *J. Mater. Res. Technol.* **2020**, *9*, 13902–13913. [CrossRef]
15. Li, W.; Tian, A.; Li, T.; Zhao, Y.; Chen, M. Ag/ZIF-8/Mg-Al LDH composite coating on MAO pretreated Mg alloy as a multi-ion-release platform to improve corrosion resistance, osteogenic activity, and photothermal antibacterial properties. *Surf. Coat. Tech.* **2023**, *464*, 129555. [CrossRef]
16. Yang, W.; Gao, Y.; Xu, D.; Zhao, J.; Ke, P.; Wang, A. Bactericidal abilities and in vitro properties of diamond-like carbon films deposited onto MAO-treated titanium. *Mater. Lett.* **2019**, *244*, 155–158. [CrossRef]
17. Chen, Q.; Zhu, X.; Jiang, Y.; Yang, L.; Liu, H.H.; Song, Z. Development and characterization of MAO/PLA-nHA nanocomposite coatings on pure zinc for orthopedic applications. *Surf. Coat. Tech.* **2024**, *478*, 130452. [CrossRef]
18. Cao, H.; Fang, M.; Jia, W.; Liu, X.; Xu, Q. Remarkable improvement of corrosion resistance of silane composite coating with $Ti_3C_2T_x$ MXene on copper. *Compos. Part B* **2022**, *228*, 109427. [CrossRef]
19. Huang, J.; Dun, Y.; Wan, Q.; Wu, Z.; Zhao, X.; Tang, Y.; Zhang, X.; Zuo, Y. Improved corrosion resistance of MAO coating on Mg-Li alloy by RGO modified silanization. *J. Alloys Compd.* **2022**, *929*, 167283. [CrossRef]
20. Han, J.; Liu, E.; Zhou, Y.; Zhao, S.; Yan, H.; Hu, C.; Kang, J.; Han, Q.; Su, Y. Robust superhydrophobic film on aluminum alloy prepared with TiO2/SiO2-silane composite film for efficient self-cleaning, anti-corrosion and anti-icing. *Mater. Today Commun.* **2023**, *34*, 105085. [CrossRef]
21. Chen, X.; Li, G.; Lian, J.; Jiang, Q. Study of the formation and growth of tannic acid based conversion coating on AZ91D magnesium alloy. *Surf. Coat. Tech.* **2009**, *204*, 736–747. [CrossRef]
22. Mandelli, A.; Bestetti, M.; Da Forno, A.; Lecis, N.; Trasatti, S.; Trueba, M. A composite coating for corrosion protection of AM60B magnesium alloy. *Surf. Coat. Tech.* **2011**, *205*, 4459–4465. [CrossRef]
23. *ASTM-B117*; Standard Practice for Operating Salt Spray (Fog) Apparatus. ASTM Standard: West Conshohocken, PA, USA, 2019.
24. Shi, C.; Zhang, L.; Zhao, J.; Tian, L.; Wang, S.; Liu, X.; Liu, G.; Shao, Y. Characterization and performance of organic-inorganic composite zinc phosphate with nano-sheet structure synthesized by a composite reaction of sol-gel with silane modification. *Surf. Interfaces* **2024**, *47*, 104225. [CrossRef]
25. Wang, G.; Guo, L.; Ruan, Y.; Zhao, G.; Zhang, X.; Liu, Y.; Kim, D.-E. Improved wear and corrosion resistance of alumina alloy by MAO and PECVD. *Surf. Coat. Tech.* **2024**, *479*, 130556. [CrossRef]
26. Yang, C.; Wang, C.; Zhao, X.; Shen, Z.; Wen, M.; Zhao, C.; Sheng, L.; Wang, Y.; Xu, D.; Zheng, Y.; et al. Superhydrophobic surface on MAO-processed AZ31B alloy with zinc phosphate nanoflower arrays for excellent corrosion resistance in salt and acidic environments. *Mater. Des.* **2024**, *239*, 112769. [CrossRef]
27. Pinto, R.; Carmezim, M.J.; Ferreira, M.G.S.; Montemor, M.F. A two-step surface treatment, combining anodisation and silanisation, for improved corrosion protection of the Mg alloy WE54. *Prog. Org. Coat.* **2010**, *69*, 143–149. [CrossRef]
28. Yang, Z.; Che, J.; Zhang, Z.; Yu, L.; Hu, M.; Sun, W.; Gao, W.; Fan, J.; Wang, L.; Liu, G. High-efficiency graphene/epoxy composite coatings with outstanding thermal conductive and anti-corrosion performance. *Compos. Part A* **2024**, *181*, 108152. [CrossRef]
29. Wang, S.; Ma, X.; Bai, J.; Niu, J.; Ma, R.; Du, A.; Zhao, X.; Fan, Y.; Li, G. Study on the structure and corrosion behavior of hot-dipped Zn–6Al–3Mg alloy coating in chlorine-containing environment. *Corros. Sci.* **2024**, *231*, 112001. [CrossRef]
30. Wang, Y.Q.; Deng, Y.Z.; Shao, Y.W.; Wang, F.H. New sealing treatment of microarc oxidation coating. *Surf. Eng.* **2014**, *30*, 31–35. [CrossRef]
31. Yang, H.; Li, S.; Liang, Z. Anodized oxidative electrosynthesis of magnesium silicate whiskers. *Int. J. Electrochem. Sci.* **2013**, *8*, 9332–9337. [CrossRef]

2. Meng, G.; Dou, B.; Zhang, T.; Liu, B.; Wang, F. Growth Behaviors of Layered Double Hydroxide on Microarc Oxidation Film and Anti-Corrosion Performances of the Composite Film. *J. Electrochem. Soc.* **2016**, *163*, C917.
3. Zanotto, F.; Grassi, V.; Frignani, A.; Zucchi, F. Protection of the AZ31 magnesium alloy with cerium modified silane coatings. *Mater. Chem. Phys.* **2011**, *129*, 1–8. [CrossRef]

Disclaimer/Publisher's Note: The statements, opinions and data contained in all publications are solely those of the individual author(s) and contributor(s) and not of MDPI and/or the editor(s). MDPI and/or the editor(s) disclaim responsibility for any injury to people or property resulting from any ideas, methods, instructions or products referred to in the content.

Article

Research on Dynamic Marine Atmospheric Corrosion Behavior of AZ31 Magnesium Alloy

Ying Wang [1,2], Weichen Xu [1,2], Xiutong Wang [1,2], Quantong Jiang [1,2], Yantao Li [1,2], Yanliang Huang [1,2] and Lihui Yang [1,2,*]

[1] Key Laboratory of Marine Environmental Corrosion and Bio-Fouling, Institute of Oceanology, Chinese Academy of Sciences, Qingdao 266071, China
[2] Open Studio for Marine Corrosion and Protection, Pilot National Laboratory for Marine Science and Technology (Qingdao), Qingdao 266237, China
* Correspondence: lhyang@qdio.ac.cn

Abstract: The dynamic marine atmospheric corrosion behavior of AZ31 magnesium alloy was investigated in situ exposed on the deck of marine scientific research vessel for 1 year. The marine scientific research vessel carried out five voyages from the coast of China to the western Pacific Ocean, while the navigation track and environmental data were collected and analyzed. The corrosion rate and characteristics were evaluated by using weight loss tests, scanning electron microscopy (SEM), X-ray diffraction (XRD), X-ray photoelectron spectroscopy (XPS), and electrochemical measurements. The corrosion rate from weight loss values was 52.23 $\mu m \cdot y^{-1}$ after exposure for 1 year, which was several times higher than that of the static field exposure test in marine atmospheric environment of other reported literature. The main corrosion products were $Mg_5(CO_3)_4(OH)_2 \cdot 4H_2O$, $MgCO_3 \cdot 3H_2O$ and $Mg_2(OH)_3Cl \cdot 4H_2O$. The corrosion was initiated from pitting corrosion and evolved into general corrosion gradually. The serious corrosion maybe due to the harsh corrosive environment with alternating changes in temperature and relative humidity caused by multiple longitude and latitude changes, and particularly high deposition rate of chloride during voyage, which was nearly twenty times that on the coast of China. This study provides effective data for the application of magnesium alloy in shipboard aircraft and other equipment, and provides a reference for indoor simulation experiments.

Keywords: magnesium alloy; dynamic marine atmospheric; corrosion; ocean voyage

Citation: Wang, Y.; Xu, W.; Wang, X.; Jiang, Q.; Li, Y.; Huang, Y.; Yang, L. Research on Dynamic Marine Atmospheric Corrosion Behavior of AZ31 Magnesium Alloy. *Metals* **2022**, *12*, 1886. https://doi.org/10.3390/met12111886

Academic Editor: Marcello Cabibbo

Received: 12 October 2022
Accepted: 2 November 2022
Published: 4 November 2022

Publisher's Note: MDPI stays neutral with regard to jurisdictional claims in published maps and institutional affiliations.

Copyright: © 2022 by the authors. Licensee MDPI, Basel, Switzerland. This article is an open access article distributed under the terms and conditions of the Creative Commons Attribution (CC BY) license (https:// creativecommons.org/licenses/by/ 4.0/).

1. Introduction

As the lightest structural metals, magnesium alloys possess good machinability and high thermal conductivity, which have been widely used in marine equipment, shipboard aircraft, and other fields [1–6]. However, magnesium alloys are susceptible to corrosion due to the high chemical and electrochemical activity, which limits its application, especially in corrosive atmospheric environments [7,8].

Many researchers have conducted a series of studies about the influence of environmental factors on magnesium alloys. Esmaily et al. [9] reported that atmospheric corrosion of Mg-Al alloy AM50 was strongly reduced with decreasing temperature. The research of Merino et al. [10] showed that corrosion attack of Mg and Mg–Al alloy under the salt fog test increased with increasing temperature. The relative humidity also affects the corrosion behavior of magnesium alloys significantly. The study of Lebozec et al. [11] showed that when the relative humidity increased from 75% to 95%, the corrosion rate of Mg–Al alloy AZ91D and AM50 increased accordingly. In addition to temperature and relative humidity, aggressive ions such as Cl⁻, accelerate the atmospheric corrosion process of magnesium alloys obviously, especially in high relative humidity environment. Jönsson et al. [12] studied the corrosion behavior of Mg–Al alloy AZ91D, which was exposed in humid air at 95% relative humidity (RH) with deposition of 70 $\mu g/cm^2$ NaCl. The results showed that the corrosion attack starts at locations with higher NaCl contents. However, most research on the atmospheric corrosion process of magnesium alloys has performed tests in simulated environment [13–19] that cannot fully simulate the synergistic effect of real atmospheric environmental factors.

Recently, some further studies on atmospheric corrosion of magnesium alloys have been conducted on the basis of exposure tests in actual atmospheric environments. Jönsson et al. [20] reported that the corrosion rate of AZ91D exposed in the marine atmospheric environment of 3–5 m from Atlantic shore Brest France was 4.2 μm/a, exposed in the rural atmospheric environment of 100 km west of Stockholm was 2.2 μm/a, and urban atmospheric environments of Stockholm was 1.8 μm/a. Liao et al. [21] found that the corrosion rate of AZ31B in the marine atmospheric environment (Shimizu, Japan) was much higher than that in urban areas (Osaka, Japan). These results indicated that magnesium alloys suffered more serious corrosion in the marine atmospheric environment.

There have been few studies of the corrosion behavior of magnesium alloys in dynamic marine atmospheric environment, and current research on atmospheric corrosion behavior of magnesium alloys were conducted with static field exposure test at permanent location, such as the coast or island. In contrast to static field exposure tests, marine equipment in application is mostly mobile in the ocean, and the harsh corrosive environment with high relative humidity, high deposition rate of chloride [22] and alternating changes in temperature and relative humidity caused by multiple longitude and latitude changes may affect the corrosion behavior of magnesium alloys. The corrosion behavior of magnesium alloy in the dynamic marine atmosphere environment of real ocean voyage has not been widely reported, and the dynamic marine atmospheric exposure experiment is a necessary complement to static exposure experiments and simulated atmospheric environments, and can provide effective data for the corrosion behavior research of magnesium alloys in the marine atmospheric environment.

In this work, the corrosion behavior of AZ31 magnesium alloys in the dynamic marine atmosphere during ocean voyage was studied through the atmospheric exposure experiment on the deck of *Research Vessel KEXUE*. In addition, the corrosion characteristics were evaluated by scanning electron microscopy (SEM), X-ray diffraction (XRD), X-ray photoelectron spectroscopy (XPS) and electrochemical measurements. This study provides effective data for the application of magnesium alloy in shipboard aircraft and other equipment.

2. Materials and Methods

2.1. Material Preparation

The specimen in this work was as-extruded AZ31 magnesium alloy, the extrusion temperature was 350 °C. The chemical composition was as listed in Table 1. Specimens for field exposure test were all 100 mm × 50 mm × 3 mm. All specimens were ground with 800 grit emery papers, degreased with acetone, dried with flowing air and weighed. Four replicate metal samples were retrieved from the exposure site after 1, 3, 6 and 12 months. Three replicas were used to determine weight loss of specimens, and the other one was used to analyze the corrosion morphology, corrosion products.

Table 1. The nominal chemical composition of AZ31 magnesium alloys (wt. %).

Material	Al	Zn	Mn	Si	Fe	Cu	Ni	Mg
AZ31	3.19	0.81	0.30	0.025	0.006	0.002	0.0006	Bal.

2.2. Dynamic Natural Environment Exposure Test

The dynamic natural environment exposure test was carried out on the open deck of the *Research Vessel KEXUE* of the Institute of Oceanology, Chinese Academy of Sciences (Qingdao, China). As is shown in Figure 1a,b, the specimens of AZ31 magnesium alloy (as circled in red in Figure 1) were installed on the test rack with the angle of 45° horizontal to the deck. The cumulative atmospheric exposure time was 1 year. As is shown in Figure 1c, the navigation range was around China offshore (Qingdao, China) to the western Pacific.

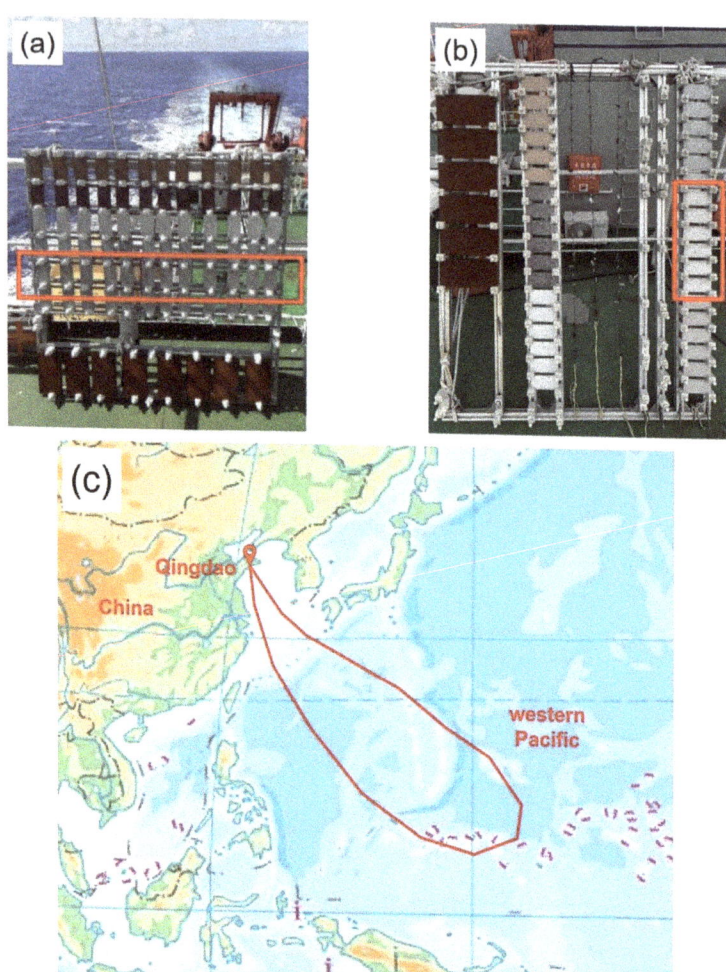

Figure 1. The dynamic natural environment exposure test environment. (**a**) Test rack A, (**b**) test rack B, (**c**) navigation range.

2.3. Determination Method for Natural Environmental Factors

The temperature, relative humidity (RH) and wind speed were measured by the automatic weather station of *Research Vessel KEXUE*.

The determination method for the deposition rate of chloride at the exposure test site described below was based on GJB 8894.1-2017. A double-layer medical gauze used to collect chloride ions with an area of 100 cm^2 was exposed at the exposure site for 7 days. Three parallel specimens of the gauze were collected each time. The collected gauze was fully cleaned, and the chloride ions concentration in the solution was measured.

2.4. Corrosion Rate Measurements

The corrosion rate was measured by weight loss measurements, and the corrosion products were removed by immersion in 200 g/L CrO_3 + 10 g/L $AgNO_3$ for 10 min at 25 °C, and then the samples were rinsed with distilled water and alcohol, dried and weighted.

The weight loss of AZ31 magnesium after exposure for different periods was calculated by using the following equation:

$$C = (w_0 - w_1)/S \tag{1}$$

where C is the weight loss of the metal due to corrosion, w_0 is the original weight, w_1 is the final weighted, S is the surface area.

The corrosion rate of AZ31 magnesium after exposure for different periods was calculated by using the following equation:

$$v = (w_0 - w_1)/(S \cdot T \cdot \rho) \tag{2}$$

where v is the corrosion rates of the metal due to corrosion, w_0 is the original weight, w_1 is the final weight, S is the surface area, T is the experimental time, ρ is the density.

2.5. Corrosion Products Analysis

Corrosion morphology of corrosion products was observed by scanning electron microscope (Regulus 8100, HITACHI, Tokyo, Japan) and Laser confocal scanning microscopy (OLS5000, Olympus, Tokyo, Japan). Phase composition was analyzed by X-ray diffraction (Ultime IV, Rigaku, Tokyo, Japan), and the element types and valence states of the corrosion products were analyzed by X-ray photoelectron spectroscopy (ESCALAB 250Xi, Thermo, Waltham, MA, USA).

2.6. Electrochemical Measurements

Electrochemical measurements were performed with a electrochemical workstation (PARSTAT 4000, Princeton Applied Research, Oak Ridge, TN, USA) in 3.5% NaCl solution in a conventional three-electrode cell, where the magnesium alloy specimen was the working electrode, saturated calomel electrode was the reference electrode and Pt foil was the counter electrode. The test system was always in a steady state with no stirring. The working electrode surface was covered with silicone rubber to leave an exposed area of 1.0 cm^2. Prior to testing, the working electrode was stabilized for about 30 min with open circuit potential measurement. Potentiodynamic polarization test was measured in the range of ± 0.5 V vs. the open circuit potential with the scan rate 1 mV/s. All the measurements were performed at ambient temperature (25 $\pm$ 2 °C) and repeated at least three times to maintain the reproducibility.

3. Results

3.1. The Environment of Field Exposure

The exposure environment of the specimens is the deck of the *Research Vessel KEXUE*, which is quite different from the static field exposure test at permanent locations such as the coast or islands reported in other studies.

Firstly, the duration of navigation in the western Pacific Ocean accounted for most of exposure time. During the exposure period, the proportion of the time of dynamic state (navigation in western Pacific Ocean) was 58.1%, and static state (stopping at Qingdao) was 41.9%. The ratio of the dynamic state and static state of exposure was about 3:2, which is more consistent with the real application environment of magnesium alloys in marine equipment.

Secondly, the *Research Vessel KEXUE* went to the western Pacific Ocean to carry out a series of scientific investigations, and traveled between Qingdao and the western Pacific five times during the exposure period, experiencing large changes in temperature during four of them. During the exposure period, the average temperature in western Pacific Ocean was about 29 °C, the average relative humidity was about 78%. In Qingdao, the annual average temperature was 14.4 °C, and the lowest temperature was below 0 °C in winter. The annual average relative humidity was 75.0%. Figure 2 shows the daily average temperature and relative humidity at exposure test site during voyage. As is shown in

Figure 2, after exposure for 3 months, the *Research Vessel KEXUE* returned to Qingdao from the western Pacific Ocean with a hot and humid environment, while at that time it was winter in Qingdao, so the temperature and relative humidity of the exposure site changed significantly. A few days later, the *Research Vessel KEXUE* went to the western Pacific, and the temperature and relative humidity rose again. The changes in temperature and relative humidity in other voyages were similar. In static exposure tests, only the change in season causes the slow change of temperature and humidity, but the specimens exposed on the deck experience rapid change of temperature and humidity several times in one year. The circulation of temperature change may cause more serious corrosion [23].

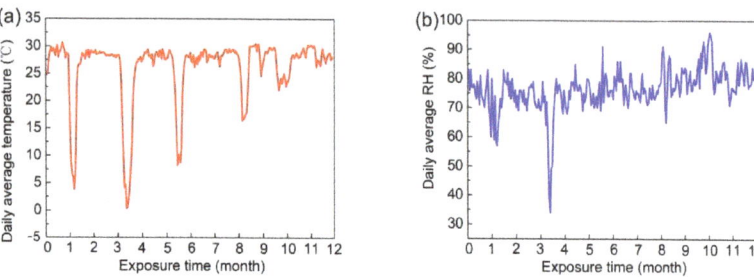

Figure 2. Daily average temperature and relative humidity at exposure test site during voyage: (**a**) temperature, (**b**) relative humidity.

Thirdly, the specimens exposed on the deck were subjected to the severe marine environment. Table 2 shows the range and average value of environment factor during the exposure period. Figure 3 shows the proportion of different range of temperature, relative humidity, deposition rate of chloride and wind speed during the ocean voyage, according to the hourly average value. Table 3 shows the range and average value of environment factor during ocean voyage. The temperature was higher than average temperature (26 °C) for most of the time, the maximum humidity was 97%, and the maximum wind speed was above 20 m/s. It is worth noting that during ocean voyage, the deposition rate of chloride was extremely high, and was above 100 mg/m^2 d most of time, and the highest value was above 1100 mg/m^2 d. The deposition rate of chloride was much higher than the value reported in other research measured in the static marine atmospheric exposure test, as shown in Table 4.

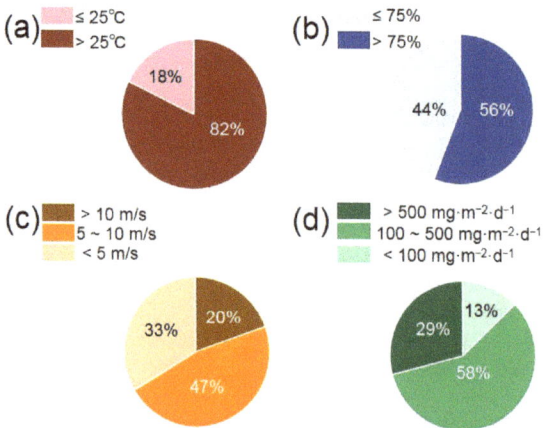

Figure 3. The proportion of environment factors during ocean voyage. (**a**) Temperature, (**b**) relative humidity, (**c**) wind speed, (**d**) deposition rate of chloride.

Table 2. The range and average value of environment factor during exposure period.

Environment Factor	T (°C)	RH (%)	Cl$^-$ (mg/m^{2}d)	Wind Speed (m/s)
Range	−0.9~33.1	18~97	63.9~1130.0	0~20.2
Average value	21.1	75.4	232.4	5.2

Table 3. The range and average value of environment factors during ocean voyage.

Environment Factor	T (°C)	RH (%)	Cl$^-$ (mg/m^{2}d)	Wind Speed (m/s)
Range	−0.9~33.1	18~97	63.9~1130.0	0.2~20.2
Average value	26.0	75.7	380.4	6.8

Table 4. The deposition rate of chloride in static marine atmospheric exposure test.

Location	Climate Type	Deposition Rate of Chloride (mg/m^{2}d)
The Gulf of Mexico [24]	subtropical monsoon	110~311
Zhanjiang, coastal of China [25]	subtropical monsoon	100~600
Xisha Islands, China [22]	tropical marine climate	64.39
Qingdao, coastal of China [26]	temperate monsoon	25
Shimizu, coastal of Japan [21]	temperate monsoon	4.2

It has been reported that chloride ions and relative humidity in the marine atmosphere significantly impact the corrosion processes of magnesium alloy [10,12,27]. Many studies illustrate the well-known corrosiveness of NaCl towards Mg alloys, and NaCl can form aqueous solution by absorbing water at RH > 75% [9]. As shown in Figure 3, the proportion of time when RH > 75% was 56%, which indicated that AZ31 magnesium alloy was covered by the thin electrolyte layer of high concentration of Cl$^-$ for more than half of the time during the ocean voyage. The thin electrolyte layer covering the surface of specimens provided the reaction environment for the electrochemical reaction during the corrosion process and made large areas on the surface become electrochemically connected.

Additionally, the chemical and electrochemical reactions involved in the anodic and cathodic reactions are thermally activated [28,29], and the effect of high temperature may also accelerate the anodic and cathodic reactions during ocean voyage.

Considering the synergistic effect of high temperature, high humidity, and high deposition rate of chloride, AZ31 magnesium alloy may suffer severe corrosion in hash dynamic marine exposure test compared with the static field exposure test at permanent location during ocean voyage.

3.2. Corrosion Rate

The weight loss of specimens exposed to the marine atmospheric environment during ocean voyage is shown in Figure 4. For the first month of the exposure period, the corrosion rate was 29.81 $\mu m \cdot y^{-1}$. However, after exposure for 3 months, the slope of the curve of weight loss increased significantly. In addition, then the weight loss of specimens increased at the similar rate with the elapse of exposure time. After exposure for 1 year, the corrosion rate was 52.23 $\mu m \cdot y^{-1}$, which was significantly higher compared with other static exposure studies. It was almost 3 times higher than that of the Xisha Islands [22] and 1.6 times higher than that of the Shimizu, Japan [21]. This means that AZ31 magnesium alloy suffered more serious corrosion in dynamic marine atmospheric environment.

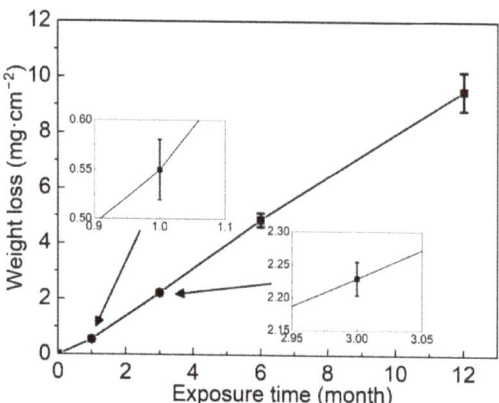

Figure 4. The weight loss of AZ31 magnesium alloy in the marine atmospheric environment during ocean voyage.

Figure 5 shows the monthly average values of temperature, relative humidity and deposition of chloride ion during exposure time. It can be seen that relative humidity and the deposition of chloride ion remained at high level. In addition, at the beginning of exposure, the deposition of chloride ion increased continuously, the maximum value appeared at the time of exposure for 3 months, almost 2 times higher than that of the first month. Therefore, the corrosion rate of AZ31 magnesium alloy increased significantly after exposure for 3 months. During the following exposure period, the AZ31 magnesium alloy was covered by thin electrolyte layer of high concentration of chloride ion in most of time under the high relative humidity and high deposition of chloride ion. Therefore, the corrosion rate of specimens remained at a high level.

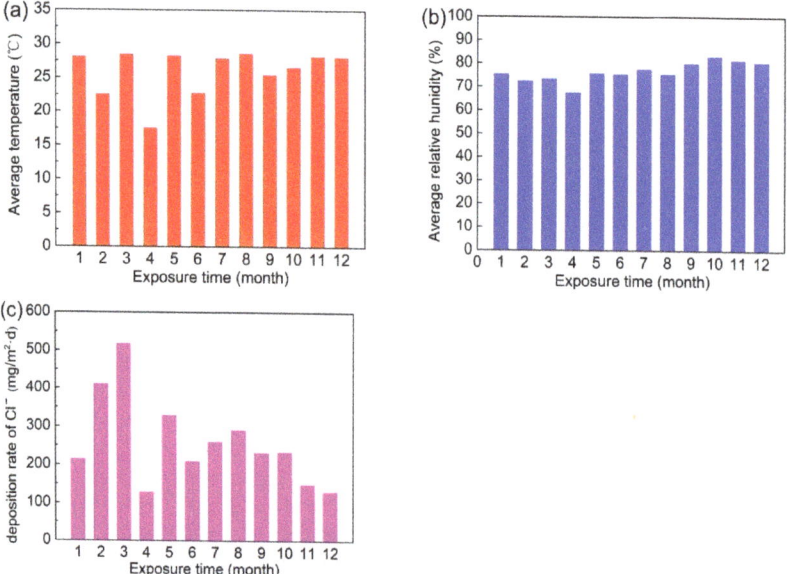

Figure 5. The monthly average values of temperature, relative humidity, and deposition of chloride ion during exposure time. (**a**) Temperature, (**b**) relative humidity and (**c**) chloride ion deposition rate.

3.3. Surface Morphology Analysis

Figure 6 shows the surface appearance of AZ31 magnesium alloy specimens with corrosion products and without corrosion products after exposure for different periods in dynamic marine environment of ocean voyage. The specimens lost their metallic luster after exposure for 1 month, and many corrosion products formed on the surface. After exposure for 12 months, the whole surface of specimens was covered by corrosion products. After removing the corrosion products, we found that the amount of corrosion pits increased, and the corrosion pits connected with each other continuously with the elapse of exposure time.

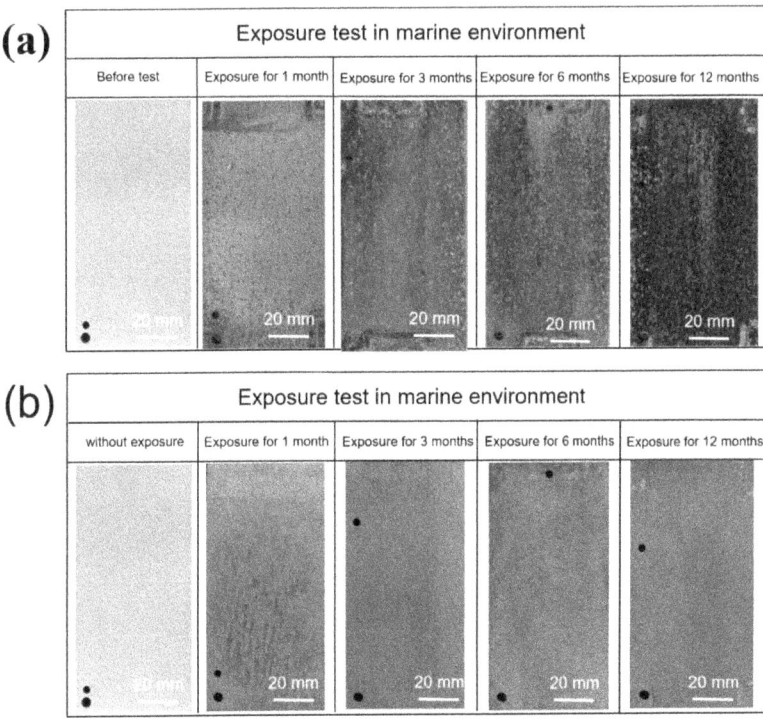

Figure 6. Surface appearance of AZ31 magnesium alloy specimens exposed for different periods in marine environment of ocean voyage: (**a**) with corrosion products, (**b**) without corrosion products.

Figure 7 displays the SEM of AZ31 magnesium alloy specimens exposed for different periods in the marine environment. A trace amount of corrosion products appeared on the surfaces of AZ31 magnesium alloy specimen after exposure for 1 month. After removing corrosion products, it can be seen that there were obvious corrosion pits on the surface of AZ31 magnesium alloy (as pointed by the arrow in Figure 7b). After exposure for 3 months, corrosion pits increased, and corrosion products completely covered the whole surface. After exposure for 6 months, a large number of corrosion products gathered on the surface of the specimens, part of surface of specimen appeared the detachment of corrosion products (as circled in red in Figure 7a), and corrosion pits connected with each other (as circled in red in Figure 7b). After exposure for 12 months, thick corrosion product layers covered the whole surface of specimen with cracks.

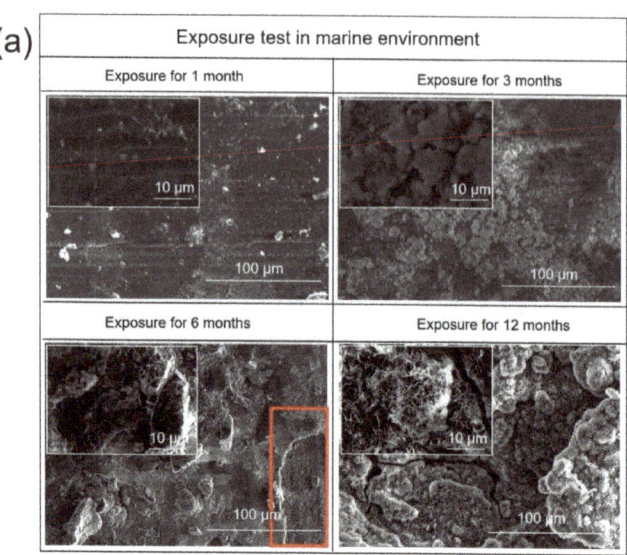

Figure 7. SEM images of the surface of AZ31 magnesium alloy specimens with corrosion products exposed for different periods in marine environment of ocean voyage: (**a**) with corrosion products, (**b**) without corrosion products.

Figure 8 shows SEM images of the cross-section of AZ31 magnesium alloy specimens exposed for 12 months in dynamic marine environment of ocean voyage. After exposure for 12 months, a corrosion product layer with a thickness of more than 50 μm was formed on the surface of specimens. However, it could also be seen that there were some small cracks in corrosion product layer. The thin electrolyte layer of high concentration of Cl^- might permeate into the matrix through these cracks, which might weaken the protection of corrosion products.

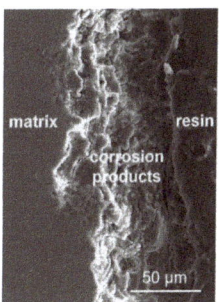

Figure 8. SEM images of the cross-section of AZ31 magnesium alloy specimens exposed for 12 months in dynamic marine environment of ocean voyage.

Figure 9 shows the laser confocal scanning microscopy (LCSM) analysis of AZ31 magnesium alloy specimens exposed for different periods in marine environment. The maximum pit depth presented a significant increase with prolonged exposure time. The maximum pit depth of specimens after exposure for 1, 3, 6 and 12 months were 44.213 μm, 63.048 μm, 172.344 μm and 276.366 μm, respectively. The research of Cui et al. [22] showed that the deepest pits of AZ31 magnesium alloy exposed on Xisha Island, with a tropical marine climate, after exposure for 1,3 and 6 months were all in the order of 30 ± 3 μm. The value of pit depth of AZ31 magnesium alloy exposed in dynamic marine atmospheric environment was significantly higher compared with other static exposure studies.

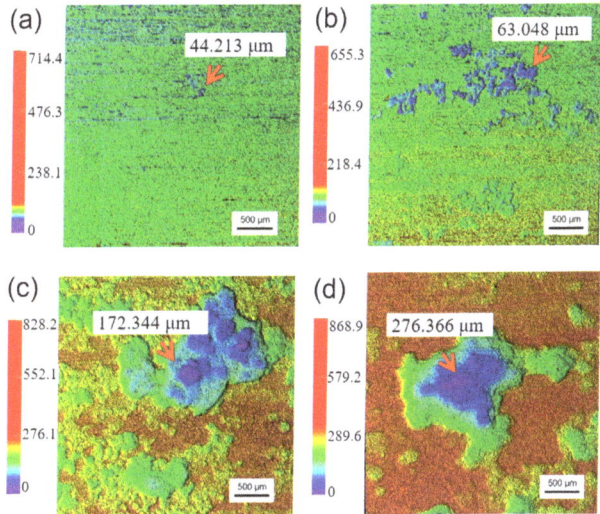

Figure 9. Laser confocal scanning microscopy (LCSM) of AZ31 magnesium alloy specimens without corrosion products exposed for different periods in marine environment of ocean voyage. (**a**) 1 month, (**b**) 3 months, (**c**) 6 months, (**d**) 12 months.

The analysis of the surface morphologies shows that the corrosion of AZ31 magnesium alloy was influenced by dynamic marine environment significantly. At the beginning of exposure, the corrosion products were formed at active sites under the corrosiveness of chloride ion. After exposure for 3 months, the average deposition rate of chloride ion was highest (in Figure 5). Under the synergistic effect of high temperature, high relative humidity and high deposition rate of chloride ion, a lot of corrosion products were formed on the surface of specimens after exposure for 3 months. After exposure for 6 months, specimens experienced temperature difference caused by ocean voyage from Qingdao to the western Pacific several times. The volume changes in the matrix and the corrosion product layer was different when temperature changed rapidly. Therefore, there was obvious stress at the interface between matrix and the corrosion product layer, which accelerated the detachment of corrosion products and the formation of cracks. The change in temperature and hash environment factor such as high wind speed and storm damaged the integrity of corrosion product layer seriously. As discussed in Section 3.1, the specimens were covered by a thin electrolyte layer of high concentration of chloride ions in most of time during ocean voyage. Therefore, the solution contained chloride ions that had stubbornly penetrated into the corrosion product layer through destroyed area of corrosion product layer, causing the amount and depth of the local corrosion to increase continually. With the extension of exposure time, more and more corrosion pits connected with each other, leading to the general corrosion and expansion of corrosion to the matrix.

3.4. Corrosion Product Analysis

The composition of corrosion products can be analyzed by XRD [30]. Figure 10 shows the composition of corrosion products formed on AZ31 magnesium alloy after exposure for 12 months. The results showed that the main corrosion products generated on AZ31 magnesium alloy were carbonate-containing compounds $Mg_5(CO_3)_4(OH)_2 \cdot 4H_2O$ (JCPDS 25-0513) [31] and $MgCO_3 \cdot 3H_2O$ (JCPDS 70-1433) [32], and chloride-containing compound $Mg_2(OH)_3Cl \cdot 4H_2O$ (JCPDS 07-0412) [33]. This indicates that CO_2 and Cl^- participated in the corrosion process.

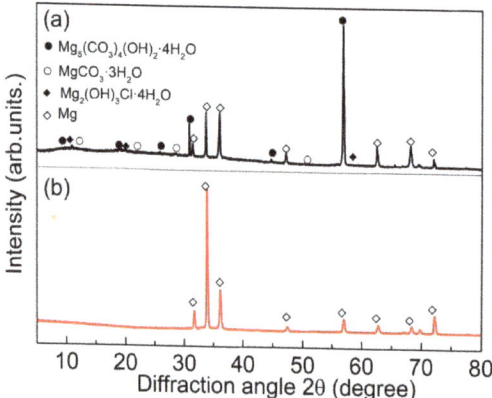

Figure 10. XRD patterns of AZ31 magnesium alloy: (**a**) corrosion products, (**b**) matrix.

Figure 11 shows the XPS spectrum of corrosion products formed on AZ31 magnesium alloy exposed for 12 months in the marine environment of ocean voyage. The element C existed as CO_3^{2-}, C-O, C=O and C-H or C-O, and carbon-containing pollutants existed on the surface of specimens. The element O existed as CO_3^{2-} and OH^-. The ratio of CO_3^{2-} and OH^- in the corrosion products was about 5.5:1, which indicated that CO_2 participated in the corrosion reaction process in the hot and humid environment and there were a large amount of CO_3^{2-} containing compounds in the corrosion products. This was consistent with the results of XRD. As shown in Table 5, the proportion of Cl 2p was 12.69%. Combined with the previous analysis of XRD and the high deposition rate of chloride, the Cl 2p in the whole-range spectra was due to chlorine-containing corrosion products and the deposition of chloride ions.

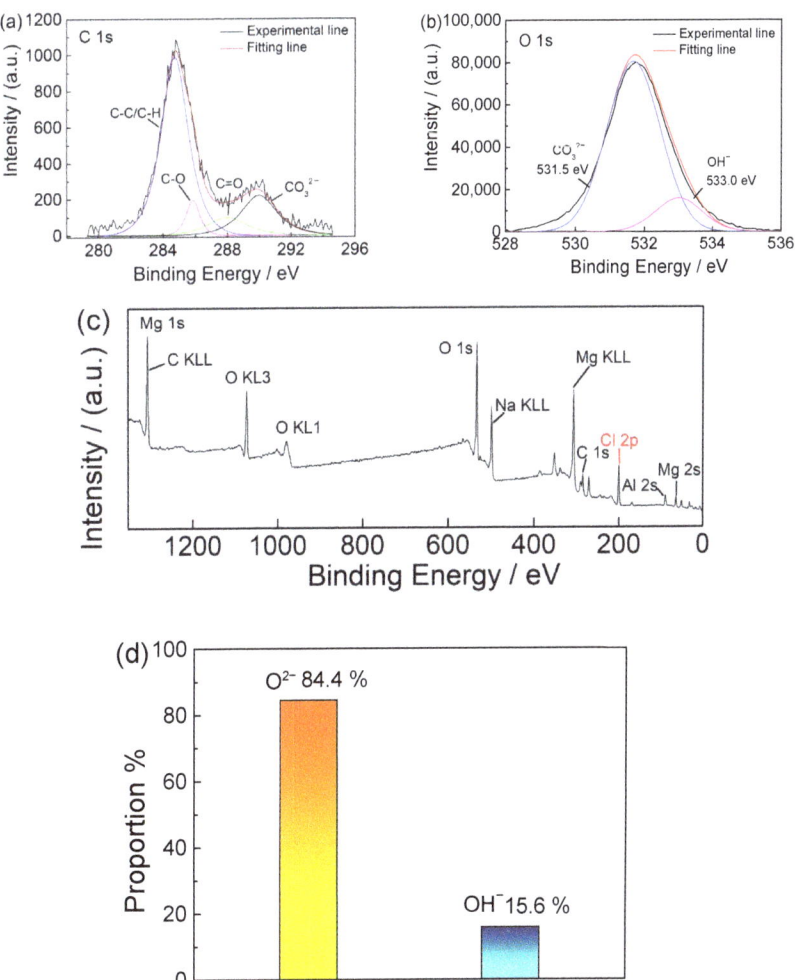

Figure 11. The XPS binding energy spectrum of corrosion products formed on AZ31 magnesium alloy exposed for 12 months in the marine environment of ocean voyage: (**a**) narrow scan spectrum of C 1s, (**b**) narrow scan spectrum of O 1s, (**c**) whole spectrum, (**d**) proportion of the different states of element O.

Table 5. The XPS analysis of the corrosion products formed on AZ31 magnesium alloy exposed for 12 months.

Element	S 2p	Cl 2p	C 1s	O 1s	Mg 2p	Zn 2p
Atomic %	1.99	12.69	29.6	41.71	13.7	0.3

3.5. Electrochemical Behavior Analysis

Figure 12 shows the polarization curves of AZ31 magnesium alloy matrix and after exposure for 12 months. The corrosion potential and current density are listed in Table 6. Compared with the AZ31 matrix, the specimens after exposed for 12 months showed a current density decreasing and a positive shift of corrosion potential. The current density decreased about one order of magnitude. This indicates that the corrosion products generated on the surface of specimens might impede the ion diffusion process [34]. With the extension of exposure, more and more corrosion products were generated on surface of the specimens under the synergistic effect of high temperature, high relative humidity and the high deposition rate of chloride ion, making the plugging effect of the corrosion product layer more obvious.

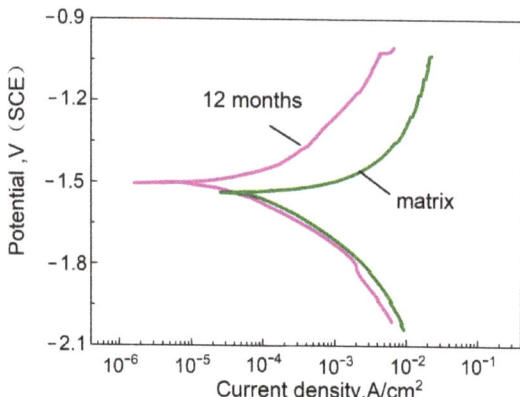

Figure 12. The polarization curves of AZ31 magnesium alloy matrix and specimen after exposure for 12 months in marine environment of ocean voyage.

Table 6. The corrosion potentials (E_{corr}) and corrosion current density (i_{corr}) obtained from anodic polarization curves of AZ31 magnesium alloy.

Specimens	Potential (E_{corr} /V)	Corrosion Current Density (i_{corr} /Acm^{-2})
Matrix	−1.532	1.054×10^{-4}
12 months	−1.508	2.984×10^{-5}

4. Discussion

Figure 13 shows the corrosion process schematic of AZ31 magnesium alloy during exposure in the dynamic marine atmosphere.

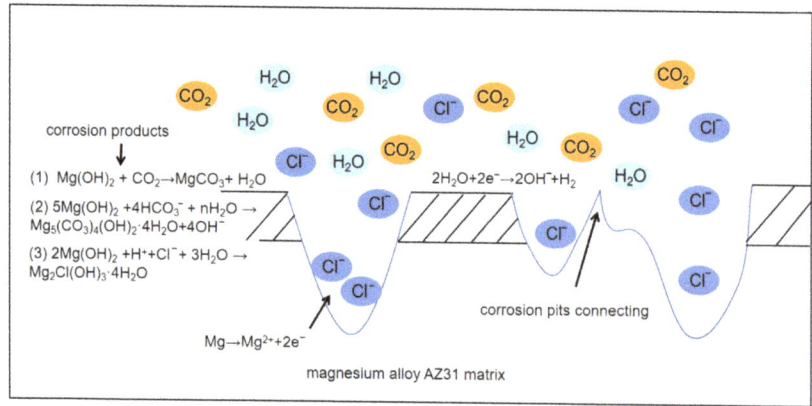

Figure 13. The corrosion process schematic of AZ31 magnesium alloy.

During the initial stage of the reaction, chloride ions attached to the defects on the specimen surface and reacted with magnesium substrate and magnesium corrosion products, and destroyed the integrity of the surface. The thin electrolyte layer of high chloride ion concentration covered on the surface of specimens provides the reaction environment for the electrochemical reaction and make large areas on the surface become electrochemically connected, the corrosion occurs rapidly, and corrosion products are generated on the surface.

The degradation of AZ31 magnesium alloy was dominated by the chemical reaction process including oxidation and hydration reactions at the beginning [9].

Anodic reaction:
$$Mg \rightarrow Mg^{2+} + 2e^- \tag{3}$$

Cathodic reaction:
$$2H_2O + 2e^- \rightarrow 2OH^- + H_2 \tag{4}$$

With the extension of exposure time, the location of local corrosion increased, Cl^- diffused into the matrix through corrosion pits, and the anodic reaction taken place inside the magnesium alloy matrix, which induced deep pits in specimens. At the same time, cathodic reaction had taken place on the specimen surface to generate OH^-, which could combine with Mg^{2+} to form magnesium hydroxide and magnesium hydroxyl-carbonate layer.

According to the phase diagram of the system $MgO / CO_2/H_2O$ [35], brucite reacted with CO_2 to form $MgCO_3$ as follows [15]:

$$Mg(OH)_2 + CO_2 \rightarrow MgCO_3 + H_2O \tag{5}$$

CO_2 reacted with H_2O to form HCO_3^-, and then reacted with brucite [22]:

$$5Mg(OH)_2 + 4HCO_3^- + nH_2O \rightarrow Mg_5(CO_3)_4(OH)_2 \cdot 4H_2O + 4OH \tag{6}$$

Brucite reacted with H^+, Cl^- and H_2O to form $Mg_2Cl(OH)_3$ as follows [36]:

$$2Mg(OH)_2 + H^+ + Cl^- + 3H_2O \rightarrow Mg_2Cl(OH)_3 \cdot 4H_2O \tag{7}$$

With the extension of exposure time, the location of local corrosion increased under the continuous action of high temperature and the thin electrolyte layer of high chloride ion concentration, the corrosion pits continuously sprout on the surface and connect with each other, the specimen evolved into general corrosion. Due to the synergistic action of the change of temperature, high wind speed and storms, corrosion products were peeled off from specimens, and all these harsh dynamic environmental factors accelerated the corrosion process.

5. Conclusions

The effect of the dynamic marine atmospheric environment on the corrosion process of AZ31 magnesium was investigated in this work. The results of this study are applicable to the coastal areas of China and the Pacific Ocean marine environment, and could provide effective data for the application of magnesium alloys in carrier plane during ocean voyage. The results can be summarized as follows:

1. The corrosion rate of AZ31 magnesium alloy after exposure for 1 year in dynamic marine atmospheric environment ocean voyage was 52.23 $\mu m \cdot y^{-1}$, the maximum depth of corrosion pits was 276.366 μm, which is much higher than that of other research studied with static field exposure test at permanent location in marine environment. After exposure for 1 year, the main corrosion products were $Mg_5(CO_3)_4(OH)_2 \cdot 4H_2O$, $MgCO_3 \cdot 3H_2O$ and $Mg_2(OH)_3Cl \cdot 4H_2O$. The corrosion was initiated from pitting corrosion and evolved into general corrosion.
2. The dynamic marine atmospheric environment is hash, with high temperature, high relative humidity and high deposition rate of chloride. The average temperature was 26.0 °C. The relative humidity was 75.7%, the proportion of time when RH > 75% was 56%. The average deposition rate of chloride ion was 380.4 $mg/m^2 d$.

3. The synergistic effect of high relative humidity and chloride ion plays an important role in the corrosion process of AZ31 magnesium alloy. During ocean voyage, the AZ31 magnesium alloy was covered by thin electrolyte layer of high concentration of chloride ion in most of time, which accelerated the corrosion of AZ31 magnesium alloy significantly.

Author Contributions: Conceptualization, L.Y.; methodology, W.X.; investigation, Q.J.; resources, L.Y.; data curation, X.W.; writing—original draft preparation, Y.W.; writing—review and editing, Y.H. and L.Y.; supervision, Y.L.; funding acquisition, L.Y. All authors have read and agreed to the published version of the manuscript.

Funding: This research was funded by the National Science and Technology Resources Investigation Program of China (Grant No. 2019FY101400), Overseas Science and education cooperation center deployment project (No. 121311KYSB20210005) and Wenhai Program of the S&T Fund of Shandong Province for Pilot National Laboratory for Marine Science and Technology (Qingdao) (NO. 2021WHZZB2304).

Institutional Review Board Statement: Not applicable.

Informed Consent Statement: Not applicable.

Data Availability Statement: Data is available from the corresponding author by request.

Acknowledgments: Data Support from Institute of Oceanology of the Chinese Academy of Sciences Marine Science Data Center (MSDC: http://msdc.qdio.ac.cn/). The data and samples were collected by RV KEXUE.

Conflicts of Interest: The authors declare no conflict of interest.

References

1. Huo, P.; Li, F.; Wu, R.; Gao, R.; Zhang, A.X. Annealing coordinates the deformation of shear band to improve the microstructure difference and simultaneously promote the strength-plasticity of composite plate. *Mater. Des.* **2022**, *219*, 110696. [CrossRef]
2. Zou, J.; Ma, L.; Jia, W.; Le, Q.; Qin, G.; Yuan, Y. Microstructural and mechanical response of ZK60 magnesium alloy subjected to radial forging. *J. Mater. Sci. Technol.* **2021**, *83*, 228–238. [CrossRef]
3. Kumar, S.D.; Kumar, S.S. Effect of heat treatment conditions on ballistic behaviour of various zones of friction stir welded magnesium alloy joints. *Trans. Nonferrous Met. Soc. China* **2021**, *31*, 156–166. [CrossRef]
4. Fu, B.; Shen, J.; Suhuddin, U.F.H.R.; Pereira, A.A.C.; Maawad, E.; Santos dos, J.F.; Klusemann, B.; Rethmeier, M. Revealing joining mechanism in refill friction stir spot welding of AZ31 magnesium alloy to galvanized DP600 steel. *Mater. Des.* **2021**, *209*, 109997. [CrossRef]
5. Wang, D.; Liu, S.; Wu, R.; Zhang, S.; Wang, Y.; Wu, H.; Zhang, J.; Hou, L. Synergistically improved damping, elastic modulus and mechanical properties of rolled Mg-8Li-4Y-2Er-2Zn-0.6Zr alloy with twins and long-period stacking ordered phase. *J. Alloys Compd.* **2021**, *881*, 160663. [CrossRef]
6. Ma, C.; Liu, Y.; Zhou, H.; Jiang, Z.; Ren, W.; He, F. Influencing mechanism of mineral admixtures on rheological properties of fresh magnesium phosphate cement. *Constr. Build. Mater.* **2021**, *288*, 123130. [CrossRef]
7. Jin, S.; Ma, X.; Wu, R.; Li, T.; Wang, J.; Krit, B.L.; Hou, L.; Zhang, J.; Wang, G. Effect of carbonate additive on the microstructure and corrosion resistance of plasma electrolytic oxidation coating on Mg–9Li–3Al alloy. *Int. J. Miner. Metall. Mater.* **2022**, *29*, 1453–1463. [CrossRef]
8. Esmaily, M.; Svensson, J.E.; Fajardo, S.; Birbilis, N.; Frankel, G.S.; Virtanen, S.; Arrabal, R.; Thomas, S.; Johansson, L.G. Fundamentals and advances in magnesium alloy corrosion. *Prog. Mater. Sci.* **2017**, *89*, 92–193. [CrossRef]
9. Esmaily, M.; Shahabi-Navid, M.; Svensson, J.-E.; Halvarsson, M.; Nyborg, L.; Cao, Y.; Johansson, L.-G. Influence of temperature on the atmospheric corrosion of the Mg-Al alloy AM50. *Corros. Sci.* **2015**, *90*, 420–433. [CrossRef]
10. Merino, M.C.; Pardo, A.; Arrabal, R.; Merino, S.; Casajús, P.; Mohedano, M. Influence of chloride ion concentration and temperature on the corrosion of Mg–Al alloys in salt fog. *Corros. Sci.* **2010**, *52*, 1696–1704. [CrossRef]
11. Lebozec, N.; Jönsson, M.; Thierry, D. Atmospheric Corrosion of Magnesium Alloys: Influence of Temperature, Relative Humidity, and Chloride Deposition. *Corrision* **2004**, *60*, 356–361. [CrossRef]
12. Jönsson, M.; Dan, P.; Thierry, D. Corrosion product formation during NaCl induced atmospheric corrosion of magnesium alloy AZ91D. *Corros. Sci.* **2007**, *49*, 1540–1558. [CrossRef]
13. Ma, X.; Jin, S.; Wu, R.; Wang, J.; Wang, G.; Krit, B.; Betsofen, S. Corrosion behavior of Mg–Li alloys: A review. *Trans. Nonferrous Met. Soc. China* **2021**, *31*, 3228–3254. [CrossRef]
14. Wan, H.; Cai, Y.; Song, D.; Li, T. Investigation of corrosion behavior of Mg-6Gd-3Y-0.4Zr alloy in Xisha atmospheric simulation solution. *Ocean Eng.* **2020**, *195*, 106760. [CrossRef]

5. Jönsson, M.; Persson, D. The influence of the microstructure on the atmospheric corrosion behaviour of magnesium alloys AZ91D and AM50. *Corros. Sci.* **2010**, *52*, 1077–1085. [CrossRef]
6. Nie, L.; Zhao, X.; Liu, P.; Hu, L.; Zhang, J.; Hu, J.; Wu, F.; Cao, F. Electrochemical behavior of ZE41 magnesium alloy under simulated atmospheric corrosion. *Int. J. Electrochem. Sci.* **2019**, *14*, 3095–3113. [CrossRef]
7. Feliu, S., Jr.; Pardo, A.; Merino, M.C.; Coy, A.E.; Viejo, F.; Arrabal, R. Correlation between the surface chemistry and the atmospheric corrosion of AZ31, AZ80 and AZ91D magnesium alloys. *Appl. Surf. Sci.* **2009**, *255*, 4102–4108. [CrossRef]
8. Lindström, R.W.; Svensson, J.-E.; Johansson, L.G. The influence of carbon dioxide on the atmospheric corrosion of some magnesium alloys in the presence of NaCl. *J. Electrochem. Soc.* **2002**, *149*, B103–B107. [CrossRef]
9. Feliu, S., Jr.; Maffiotte, C.; Galván, J.C.; Barranco, V. Atmospheric corrosion of magnesium alloys AZ31 and AZ61 under continuous condensation conditions. *Corros. Sci.* **2011**, *53*, 1865–1872. [CrossRef]
10. Jönsson, M.; Persson, D.; Leygraf, C. Atmospheric corrosion of field-exposed magnesium alloy AZ91D. *Corros. Sci.* **2008**, *50*, 1406–1413. [CrossRef]
11. Liao, J.; Hotta, M.; Motoda, S.; Shinohara, T. Atmospheric corrosion of two field-exposed AZ31B magnesium alloys with different grain size. *Corros. Sci.* **2013**, *71*, 53–61. [CrossRef]
12. Cui, Z.; Li, X.; Xiao, K.; Dong, C. Atmospheric corrosion of field-exposed AZ31 magnesium in a tropical marine environment. *Corros. Sci.* **2013**, *76*, 243–256. [CrossRef]
13. Man, C.; Dong, C.; Wang, L.; Kong, D.; Li, X. Long-term corrosion kinetics and mechanism of magnesium alloy AZ31 exposed to a dry tropical desert environment. *Corros. Sci.* **2020**, *163*, 108274. [CrossRef]
14. Soares, G.C.; Garbatov, Y.; Zayed, A.; Wang, G. Influence of environmental factors on corrosion of ship structures in marine atmosphere. *Corros. Sci.* **2009**, *51*, 2014–2016. [CrossRef]
15. Chen, H.; Cui, H.; He, Z.; Lu, L.; Huang, Y. Influence of chloride deposition rate on rust layer protectiveness and corrosion severity of mild steel in tropical coastal atmosphere. *Mater. Chem. Phys.* **2021**, *259*, 123971. [CrossRef]
16. Sun, X.; Lin, P.; Man, C.; Cui, J.; Wang, H.; Dong, C.; Li, X. Prediction model for atmospheric corrosion of 7005-T4 aluminum alloy in industrial and marine environments. *Int. J. Miner. Metall. Mater.* **2018**, *25*, 1313–1319. [CrossRef]
17. Lindström, R.; Johansson, L.; Thompson, G.E.; Skeldon, P.; Svensson, J. Corrosion of magnesium in humid air. *Corros. Sci.* **2004**, *46*, 1141–1158. [CrossRef]
18. Niklasson, A.; Johansson, L.; Svensson, J. The influence of relative humidity and temperature on the acetic acid vapour-induced atmospheric corrosion of lead. *Corros. Sci.* **2008**, *50*, 3031–3037. [CrossRef]
19. Nazir, M.H.; Saeed, A.; Khan, Z. A comprehensive predictive corrosion model incorporating varying environmental gas pollutants applied to wider steel applications. *Mater. Chem. Phys.* **2017**, *193*, 19–34. [CrossRef]
20. Kumar, P.; Pathak, S.; Singh, A.; Jain, K.; Khanduri, H.; Wang, L.; Kim, S.K.; Pant, R.P. Observation of intrinsic fluorescence in cobalt ferrite magnetic nanoparticles by Mn^{2+} substitution and tuning the spin dynamics by cation distribution. *J. Mater. Chem. C* **2022**, *10*, 12652–12679. [CrossRef]
21. Pang, H.; Tian, P.; Wang, J.; Wang, X.; Ning, G.; Lin, Y. Fabrication of microstructured $Mg_5(CO_3)_4(OH)_2$ $4H_2O$ and $MgCO_3$ in flue gas absorption technology. *Mater. Lett.* **2014**, *131*, 206–209. [CrossRef]
22. Wang, Y.; Li, Z.; Demopoulos, G.P. Controlled precipitation of nesquehonite ($MgCO_3 \cdot 3H_2O$) by the reaction of $MgCl_2$ with $(NH_4)_2CO_3$. *J. Cryst. Growth* **2008**, *310*, 1220–1227. [CrossRef]
23. Lojka, M.; Jiříčková, A.; Lauermannová, A.M.; Pavlíková, M.; Pavlík, Z.; Jankovský, O. Kinetics of formation and thermal stability of $Mg_2(OH)_3Cl \cdot 4H_2O$. *AIP Conf. Proc.* **2019**, *2170*, 020009.
24. Kim, S.H.; Yeon, S.; Lee, J.H.; Kim, Y.W.; Lee, H.; Park, J.; Lee, N.; Choi, J.P.; Aranas, C., Jr.; Lee, Y.J.; et al. Additive manufacturing of a shift block via laser powder bed fusion: The simultaneous utilisation of optimised topology and a lattice structure. *Virtual Phys. Prototyping* **2020**, *15*, 460–480. [CrossRef]
25. Zhang, Z.; Fedortchouk, Y.; Hanley, J.J.; Kerr, M. Diamond resorption and immiscibility of C-O-H fluid in kimberlites: Evidence from experiments in $H_2O—CO_2—SiO_2—MgO—CaO$ system at 1–3 GPa. *Lithos* **2021**, *380–381*, 105858. [CrossRef]
26. Sun, L.; Dong, N.; Wang, J.; Ma, H.; Jin, P.; Peng, Y. Effect of solid solution Zn atoms on corrosion behaviors of Mg-2Nd-2Zn alloys. *Corros. Sci.* **2022**, *196*, 110023. [CrossRef]

Article

Simultaneous Improvement of Strength, Ductility and Damping Capacity of Single β-Phase Mg–Li–Al–Zn Alloys

Xinhe Yang [1,†], Yang Jin [1,†], Ruizhi Wu [1,*], Jiahao Wang [1,2,*], Dan Wang [3], Xiaochun Ma [1], Legan Hou [1], Vladimir Serebryany [4], Iya I. Tashlykova-Bushkevich [5] and Sergey Ya. Betsofen [6]

1. Key Laboratory of Superlight Materials & Surface Technology, Ministry of Education, Harbin Engineering University, Harbin 150001, China
2. Jiangxi Key Laboratory of Forming and Joining Technology for Aerospace Components, Nanchang Hangkong University, Nanchang 330063, China
3. Key Laboratory of New Carbon-Based Functional and Super-Hard Materials of Heilongjiang Province, School of Physics and Electronic Engineering, Mudanjiang Normal University, Mudanjiang 157011, China
4. Baikov Institute of Metallurgy and Materials Science, Russian Academy of Sciences, Moscow 119334, Russia
5. Physics Department, Belarusian State University of Informatics and Radioelectronics, P.Brovki Str. 6, 220013 Minsk, Belarus
6. Moscow Aviation Institute (MAI), National Research University, Moscow 125993, Russia
* Correspondence: rzwu@hrbeu.edu.cn (R.W.); wjhyjyo@163.com (J.W.)
† These authors contributed equally to this work.

Citation: Yang, X.; Jin, Y.; Wu, R.; Wang, J.; Wang, D.; Ma, X.; Hou, L.; Serebryany, V.; Tashlykova-Bushkevich, I.I.; Betsofen, S.Y. Simultaneous Improvement of Strength, Ductility and Damping Capacity of Single β-Phase Mg–Li–Al–Zn Alloys. Metals 2023, 13, 159. https://doi.org/10.3390/met13010159

Academic Editor: Talal Al-Samman

Received: 10 December 2022
Revised: 9 January 2023
Accepted: 10 January 2023
Published: 12 January 2023

Copyright: © 2023 by the authors. Licensee MDPI, Basel, Switzerland. This article is an open access article distributed under the terms and conditions of the Creative Commons Attribution (CC BY) license (https://creativecommons.org/licenses/by/4.0/).

Abstract: Body-centered cubic (BCC) Mg–Li alloy can be effectively strengthened by with the addition of Al and Zn. However, adding excessive amounts result in reduced mechanical properties and damping capacity of the alloy during subsequent heat treatment and deformation. The effects of solution-hot rolling-aging on the mechanical properties and damping capacity of LAZ1333 alloy and LAZ1366 alloy were studied. The solid solution strengthening greatly increases the hardness of the alloy, but the ductility is extremely poor. The AlLi softening phase precipitated during the subsequent hot rolling and aging process greatly improves the ductility of the alloy, but the excess precipitation of in the AlLi softening phase and the solid solution of excess Zn element are not conducive to the substantial improvement of the strength and ductility of the alloy. Excessive addition of alloying elements is detrimental to the damping capacity of the alloy, but the damping capacity of the alloy can be significantly improved by depleting the number of solute atoms through subsequent ageing treatments. The UTS and FE of as-cast LAZ 1333 alloy are 111 MPa and 16.9%, respectively. The as-aged LAZ1333 alloy has the best mechanical properties and damping capacity, and the UTS and FE are increased by 65.8% and 89.3%, respectively, compared to the as-cast alloy, and the damping capacity increased from 0.011 to 0.015.

Keywords: BCC-structured Mg–Li alloy; mechanical properties; damping capacity; aging softening; dislocation strengthening; movable dislocation density

1. Introduction

Mg–Li alloys are widely used in aerospace applications, medical devices, weapons, automobiles and 3C industries due to their low density, high specific stiffness and specific strength, excellent electromagnetic shielding properties and damping properties [1–3]. Based on the Mg–Li binary phase diagram, Mg–Li alloy is completely composed of body-centered cubic (BCC) β-Li single phase when the Li content exceeds 10.3 wt%. Numerous studies have shown that BCC β-Li single phase Mg–Li alloys have excellent ductility but low strength and damping capacity. The disadvantage of low strength and poor damping capacity have limited the further application of Mg–Li alloys to some extent. Generally, the damping capacity and mechanical properties of Mg alloys are an ambivalence, which is confirmed by the available literature [4–6]. At present, the disadvantage of low strength and poor damping capacity of BCC β-Li single phase Mg–Li alloy can be effectively solved by alloying, heat treatment and plastic deformation. Among them, Al and Zn elements are common alloying elements that can effectively strengthen Mg–Li alloys. The strengthening of the Mg–Li alloy by the Al element is mainly through solid solution strengthening and precipitation strengthening. The generated $MgLi_2Al$ belongs to the strengthening phase, and the transition of metastable phases ($MgLi_2Al \rightarrow AlLi$) is thought to result in age softening, which is decomposed into the AlLi softening phase during the aging process [7]. The Zn element has little damage to the plasticity of the Mg–Li alloy, and it mainly plays the role of solid-solution strengthening in the alloy. Al and Zn elements are added as alloying elements at the same time, which not only gives full play to the advantages of Al and Zn elements, but it also avoids the negative effects of excessive single alloying elements [8–11]. In addition, heat treatments (e.g., solid solution and aging) are effective methods to optimize the microstructure, mechanical properties and damping capacity of Mg–Li alloys. By precipitation of fine second phases to deplete the solute atoms in the matrix, it leads to an increase in dislocation movement distances and thus effectively improves the damping capacity of the alloy. Plastic deformation methods such as rolling can also effectively improve the mechanical properties of Mg–Li alloys through deformation strengthening [12–14]. Liu rolled the Mg–8Li–3Al–2Zn–0.5Y alloy at 200 °C, and found that the grain was elongated, and the tensile strength increased from 200 MPa to 250 MPa with the increase of deformation. Ji rolled the squeezed the Mg–16Li–2.5Zn–2.5Er alloy at room temperature, and observed a large number of dislocations in the alloy. The tensile strength of the alloy increased from 93 MPa as cast to 234 MPa as rolled, and the alloy obtained a high specific strength of 178 kN·m/kg.

In order to prepare Mg–Li matrix alloy with low density and excellent properties, Mg–13Li–xAl–yZn (x = 3, 6 wt%, y = 3, 6 wt%) alloys was selected as the research object to study the effects of heat treatment and rolling deformation on the mechanical properties and damping capacity of the alloy. Existing studies show that the mechanical properties of the Mg–Li alloy gradually improve with the increase of aluminum content. When the aluminum content is greater than 5–6%, the alloy strength does not improve significantly, but the elongation decreases significantly. Therefore, the aluminum content in Mg–Li–Al alloy is generally lower than 5–6%. When the aluminum content reaches more than 3%, the aluminum will not only dissolve in the alloy matrix, but also appear AlLi phase. Zinc has a high solid solubility (about 6.2%) in magnesium, and with the decrease of temperature, the solid solubility decreases, resulting in aging strengthening. In the process of magnesium lithium alloy alloying, aluminum and zinc are generally added as alloying elements at the same time, so as to give full play to their respective advantages. Therefore, the alloy content for both Zn and Al as 3 and 6 wt%.

2. Methods

In this study, Mg–13Li–xAl–yZn (x = 3, 6 wt%, y = 3, 6 wt%) (LAZ13xy) alloys were prepared by melting pure magnesium (99.95 wt.%), pure lithium (99.90 wt.%), pure aluminum (99.90 wt.%) and pure zinc (99.90 wt.%) in an induction melting furnace under ambient argon gas. The melt was poured into a steel mold to obtain as-cast specimens. The as-cast alloys were solution treated at 400 °C for 5 h followed by rapid air cooling, and the obtained sample was called the as-solutionized alloy. Subsequently, the solid-solution alloy was rolled at 200 °C at a rolling speed of 600 cm/min. The reduction amount of each pass was 0.2 mm and the total reduction rate was 50% to get the hot-rolled alloy. During the whole rolling process, the roll is not preheated and no lubricant is used. The hot-rolled alloy was then aged at 80 °C for 12 h.

The microstructural observation was performed by optical microscope (OM, LEICA DM-IRM, Leica Microsystems, Wetzlar, Germany), scanning electron microscope (SEM, SU5000, Hitachi High-tech Company, Tokyo, Japan) and transmission electron microscope (TEM, FEI Talos F200X G2, Thermo Fisher Scientific, Waltham, MA, USA). The sample was polished with sandpaper after optical microscope observation, and then corroded with

1% nitrate alcohol for 5 s. After quickly cleaning the sample with alcohol, the surface of the sample was dried. The phase composition of the specimens was measured by X-ray diffraction (XRD, Rigaku TTR-III, Rigaku Corporation, Tokyo, Japan) with Cu-Kα radiation at a scanning rate of 5°/min between 20° and 80°. The tensile fracture morphologies of the specimens were observed by scanning electron microscope (SEM, JSM-6480A, JEOL, Tokyo, Japan).

The hardness of LAZ13xy (x = 3, 6 wt%, y = 3, 6 wt%) alloys was measured by Vickers hardness testing with a load of 200 gf and a duration of 15 s. For each specimen, at least eight indents were performed, and the average values were used. The strength and elongation of these alloys were measured with tensile tester (1 mm/min of the tensile rate). The average values were obtained from three parallel tensile tests. The damping capacity was measured by a dynamic mechanical analyzer in single-cantilever vibration mode (TA-DMA Q800, New Castle, DE, USA), and the vibration frequency is 1 Hz. The dimensions of the specimen are 35 mm × 10 mm × 1 mm.

3. Result and Discussion

3.1. Microstructures and Hardness

Figure 1 manifests the microstructures of as-cast LAZ13xy (x = 3, 6 wt%, y = 3, 6 wt%) alloys. The grain size of the as-cast LAZ1333 alloy are coarse, with an average size of 223 µm. In addition, a small amount of second-phase particles precipitated inside the grains and at the grain boundaries. Compared with the as-cast LAZ1333 alloy, the as-cast LAZ1366 alloy has smaller grain size, with an average grain size of 153 µm, and the amount of second phase precipitated inside the grains and at the grain boundaries is significantly increased. The main reason for the fine grain size of the as-cast LAZ1366 alloy is that the second phase precipitated along the grain boundary during the solidification plays a pinning role, thus hindering the growth of the grains. Figure 2 shows that the as-cast LAZ1366 alloy has more second phases compared to the as-cast LAZ1333 alloy. In order to analyze the second phases, EDS and XRD was carried out on the as-cast LAZ1333 alloy and LAZ1366 alloy. Table 1 shows the EDS results of the second phases marked by the yellow arrows and letters in Figure 2. Combined with the XRD results (Figure 3), we can conclude that the as-cast LAZ1333 alloy and LAZ1366 alloy are composed of β-Li and AlLi phases.

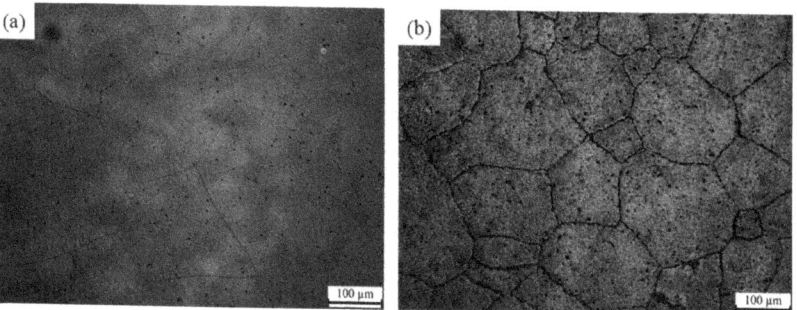

Figure 1. The OM images of as-cast LAZ13xy (x = 3, 6 wt%, y = 3, 6 wt%) alloys: (**a**) LAZ1333, (**b**) LAZ1366.

Table 1. The EDS results of the second phases marked by the yellow arrows and letters in Figure 2.

Positions	Chemical Compositions (at%)		
	Mg	Al	Zn
A	87.03	9.10	3.87
B	87.20	9.44	3.36
C	90.43	6.79	2.78
D	93.36	4.50	2.14

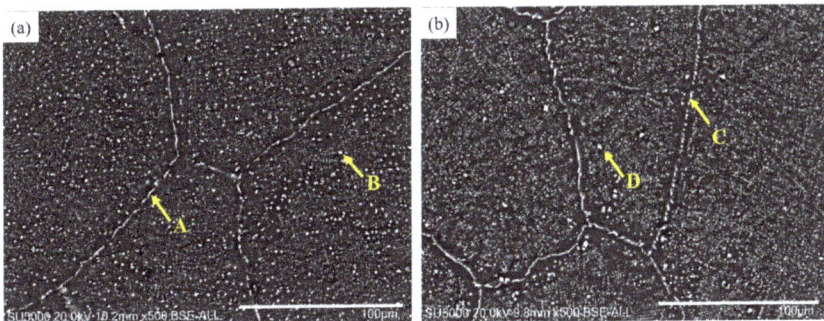

Figure 2. The SEM images of as-cast LAZ13xy (x = 3, 6 wt%, y = 3, 6 wt%) alloys: (**a**) LAZ1333, (**b**) LAZ1366.

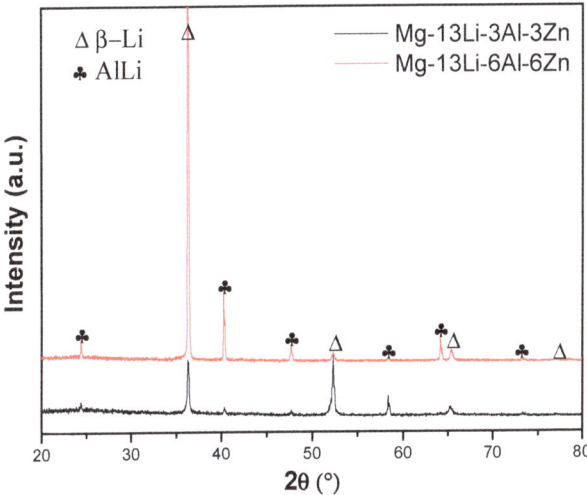

Figure 3. The XRD patterns of as-cast LAZ13xy (x = 3, 6 wt%, y = 3, 6 wt%) alloys.

The as-cast LAZ1366 alloy was then subjected to TEM analysis. As shown in Figure 4a, the precipitation phase in the alloy is micrometer-scale, and the spinodal decomposition appears around the precipitation phase. The HRTEM characterization of the spinodal decomposition is shown in Figure 4b. Figure 4c is the FFT result of region 1, which shows a typical BCC[001] axis, which is β-Li matrix. Figure 4d shows the FFT result of the spinodal decomposition in region 2, which shows a typical BCC[$\bar{1}$22] axis. According to the HAADF-STEM and EDS results shown in Figure 5, these micrometer-scale precipitates hardly contain Mg elements, and only contain Al and Zn elements. It can be seen from the EDS results that the atomic ratio of Al and Zn elements in the precipitation phase is close to 3:1. Combining with the XRD results (Figure 3) and related literature [15–17], the micrometer-scale precipitation phase is AlLi phase. However, no Zn-containing compounds were found in LAZ1333 alloy and LAZ1366 alloy, which was mainly because the maximum solid solubility of Zn in Mg is about 6.2 wt.%, and the Zn element was completely dissolved in the matrix [18,19].

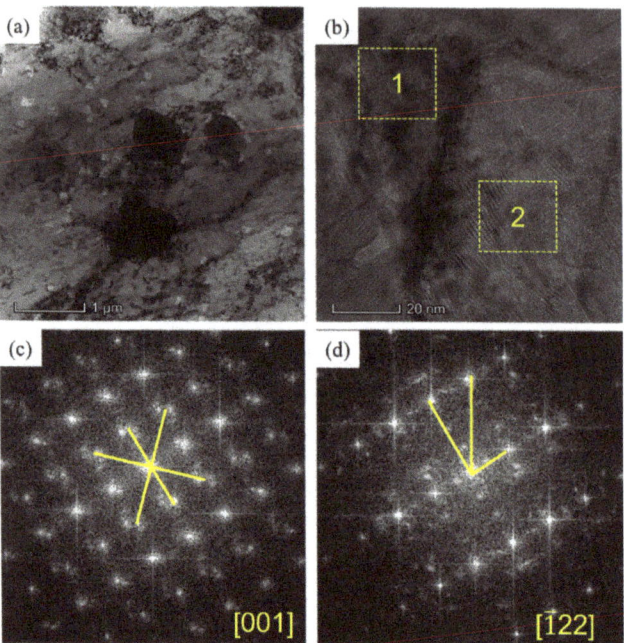

Figure 4. (**a**) TEM image and (**b**) HRTEM image of spinodal decomposition organization in the as-cast LAZ1366 alloy; FFT results of (**c**) area 1 and (**d**) area 2 in (**b**).

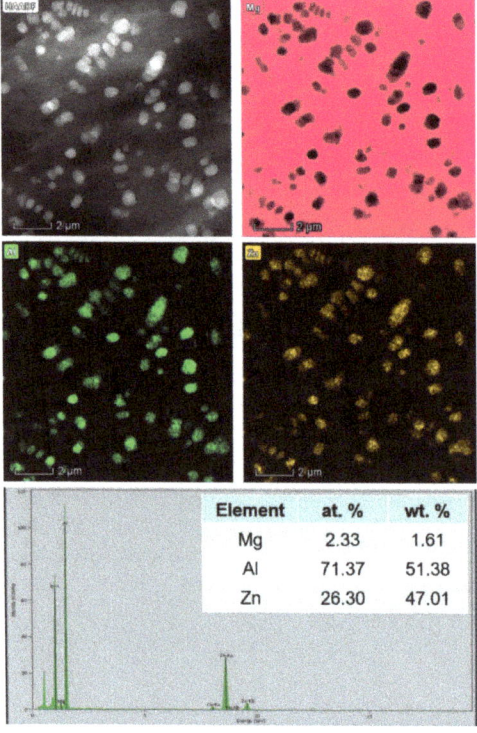

Figure 5. HADDF-STEM image of as-cast LAZ1366 alloy and its elemental mapping.

The as-cast alloy was subjected to solution treatment at 400 °C for 5 h. Figure 6a,b show the microstructures of the as-solutionizied LAZ13xy (x = 3, 6 wt%, y = 3, 6 wt%) alloys. It can be seen that the second phase inside the grains and at the grain boundaries in the as-cast alloy disappears, and the average grain sizes of the alloys are 187 μm and 180 μm, respectively. Then the XRD analysis of the as-solutionizied alloy was carried out. From Figure 6c, both the as-solutionized LAZ1333 alloy and the as-solutionized LAZ1366 alloy only have the diffraction peaks of the β-Li phase, which indicates that the AlLi phase is completely dissolved into the matrix. As shown in Figure 7, the hardness of as-cast LAZ1333 alloy and LAZ1366 alloy are 45.2 Hv and 57 Hv, respectively. After solution treatment, the hardness values are increased to 110.6 Hv and 121.3 Hv, respectively. The significant increase in the hardness of the as-solutionized alloys is caused by the decomposition of the AlLi softening phase and the solid solution of Al and Zn elements. The heat flux stress of alloy can be calculated by Al content. The higher the Al content, the higher the heat flux stress, and the higher the creep resistance may be. Therefore, the hardness of as-cast and solid-solution LAZ1366 alloy is higher [20].

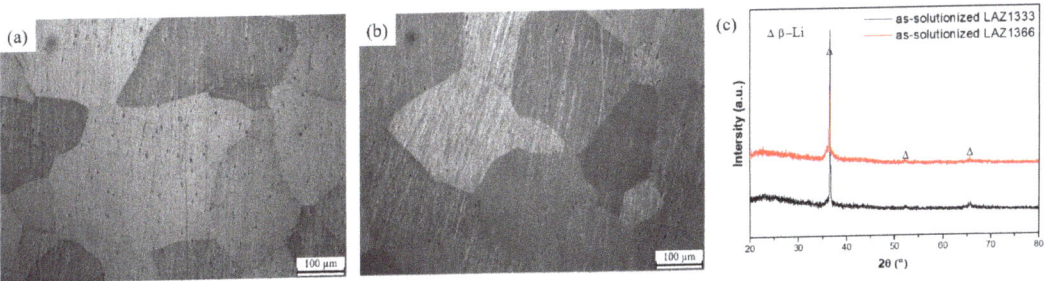

Figure 6. The microstructure of as-solutionized LAZ13xy (x = 3, 6 wt%, y = 3, 6 wt%) alloys: OM images of (**a**) LAZ1333 and (**b**) LAZ1366, (**c**) XRD pattern.

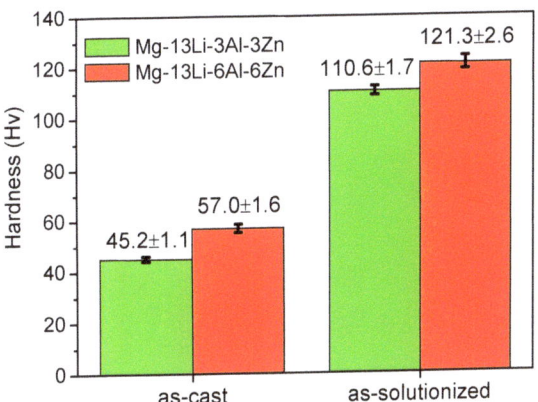

Figure 7. Hardness histograms for as-cast and as-solutionized LAZ13xy (x = 3, 6 wt%, y = 3, 6 wt%) alloys.

The as-solutionized alloy was hot rolled at 200 °C, and the phase analysis of the alloy is shown in Figure 8a. The hot rolling resulted in dynamic precipitation of the AlLi phase. In general, the dynamic precipitation depends on the diffusion rate of atoms during thermal deformation [21]. Plastic deformation leads to a large number of dislocations and vacancies, which can provide additional channels for atomic diffusion [22]. Secondly, the solid solution treatment was carried out at 400 °C, but the rolling temperature was 200 °C. The decrease in temperature leads to a decrease in the solubility of the Al and Zn atoms, which stimulates the precipitation of solute atoms from the supersaturated matrix in the form of a second phase [23–25]. It can be seen from Figure 8b that the hardness values of the hot-rolled LAZ1333 alloy and hot-rolled LAZ1366 alloy are reduced to 85.3 Hv and 71 Hv, respectively. It is worth noting that the hardness of hot-rolled LAZ1366 alloy greatly decreases, which is mainly caused by the excessive precipitation of the AlLi phase. As shown in Figure 8b, the hardness of LAZ1333 alloy reached the lowest point at 1 h of aging, and then gradually increases, reaching equilibrium after 4 h with a value of 74 Hv. The hardness of LAZ1366 alloy gradually decreases with the increase of aging time, and the hardness is 65.4 Hv when aging for 4 h. Figure 9a,b show the SEM images of the as-aged (80 °C—4 h) LAZ13xy (x = 3, 6 wt%, y = 3, 6 wt%) alloy. It can be seen that many granular second phases are precipitated in the LAZ1333 and LAZ1366 alloys, and there are more granular second phases and larger sizes in the LAZ1366 alloy. Table 2 shows the EDS results of the second phases marked by the yellow arrows and letters in Figure 9 a,b. From the XRD pattern results in Figure 9c, it can be seen that the as-aged LAZ1333 alloy and LAZ1366 alloy only have the diffraction peaks of the β-Li phase and AlLi phase. The decrease in the hardness of the as-aged alloy is caused by the precipitation of the AlLi phase consuming the Al atoms in the matrix.

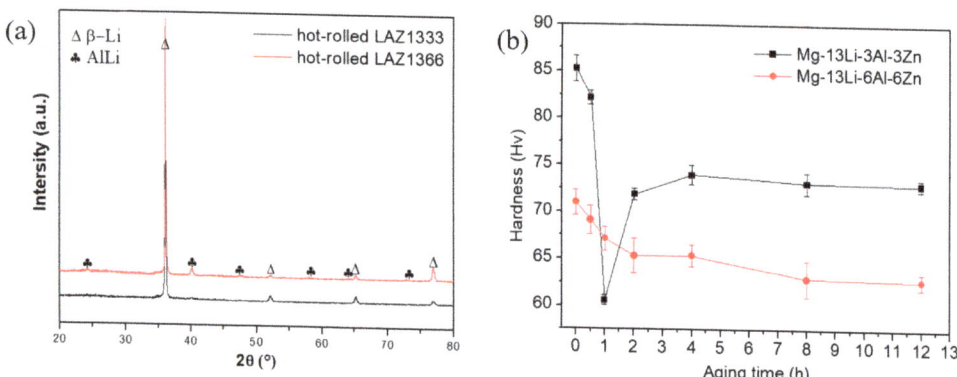

Figure 8. (**a**) XRD pattern of hot-rolled LAZ13xy (x = 3, 6 wt%, y = 3, 6 wt%) alloys, (**b**) aging hardening curves of the LAZ13xy (x = 3, 6 wt%, y = 3, 6 wt%) alloys at 100 °C.

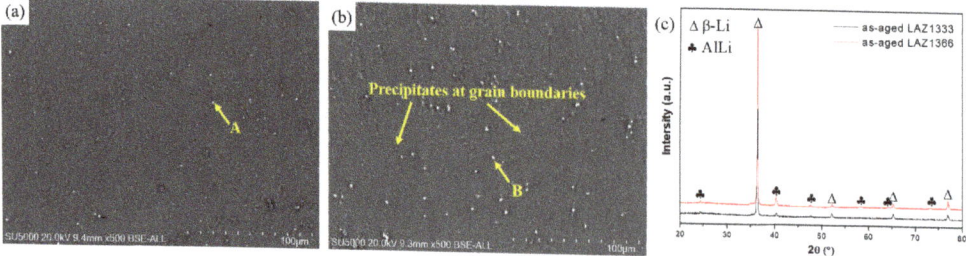

Figure 9. The microstructure of as-aged LAZ13xy (x = 3, 6 wt%, y = 3, 6 wt%) alloys: SEM images of (**a**) LAZ1333 and (**b**) LAZ1366, (**c**) XRD pattern.

Table 2. The EDS results of the second phases marked by the yellow arrows and letters in Figure 9a,b.

Positions	Chemical Compositions (at%)		
	Mg	Al	Zn
A	72.24	25.94	1.82
B	54.08	43.95	1.97

3.2. Mechanical Properties

Figure 10 shows the tensile stress–strain curves for as-cast LAZ1333 and LAZ1366, coupled with SEM fracture morphology. The ultimate tensile strength (UTS) of the as-cast LAZ1333 and LAZ1366 are only 111 MPa and 164 MPa, while the fracture elongations (FE) are 16.9% and 13.4%. The presence of excessive Al and Zn elements in the matrix gives LAZ1366 alloys higher strength. The overall fracture characterization of as-cast LAZ1333 alloy exhibits typical ductile fracture with many large and deep dimples. The fracture surface of as-cast LAZ1366 alloy is composed of tearing ridges and small amounts of dimples, indicating a mixed fracture of brittle and ductile. Figure 1b shows that there are too many precipitates precipitated along grain boundaries in LAZ1366 alloy, which greatly hinders the dislocations initiating and slipping within these zones. Therefore, the FE of as-cast LAZ1366 alloy is lower than that of as-cast LAZ1333 alloy. However, a complete tensile curve cannot be obtained due to the extremely poor plasticity of the as-solutionized alloy. From the SEM fracture morphology of Figure 10, it can be seen that the overall fracture characteristics of the as-solutionized alloy show intergranular fracture with some cleavage fractures. Some cleavage facets contain a number of shallow river markings. Typically, the river markings were expanded to the entire grain from a small zone and are parallel to each other. However, some of the river markings are radial or irregular (blue arrow grain in Figure 11), which could be ascribed to the cracking originated from local grain boundary decohesion [21].

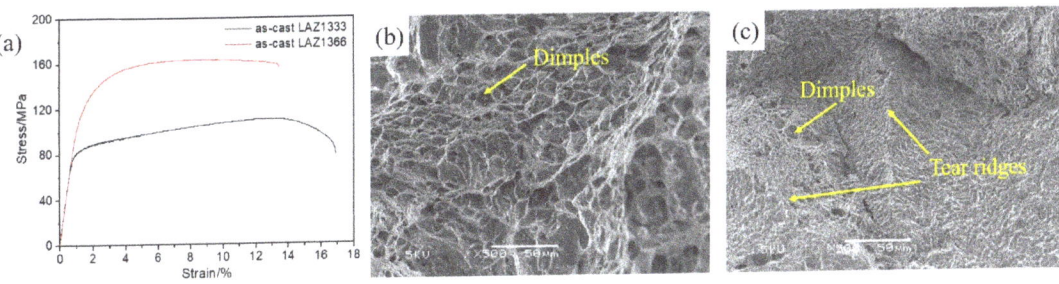

Figure 10. Room temperature tensile stress–strain curves and fracture morphology of the as-cast LAZ13xy (x = 3, 6 wt%, y = 3, 6 wt%) alloys: (**a**) tensile stress–strain curves, fracture morphology of as-cast (**b**) LAZ1333 and (**c**) LAZ1366.

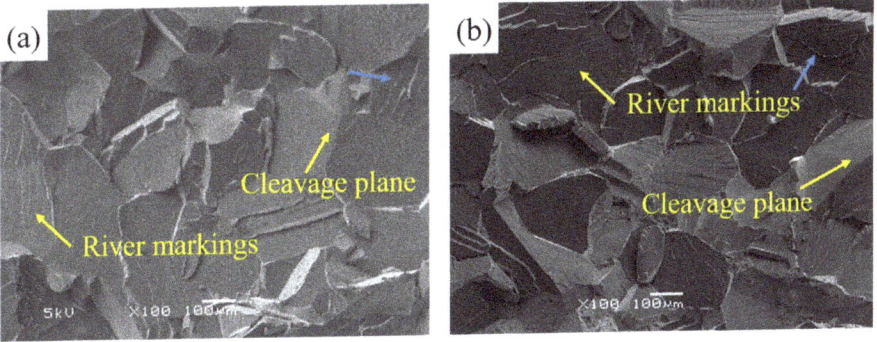

Figure 11. Tensile fracture morphology of the as-solutionized LAZ13xy (x = 3, 6 wt%, y = 3, 6 wt%) alloys: (**a**) LAZ1333, (**b**) LAZ1366.

Figure 12 shows the tensile stress–strain curves for as-rolled LAZ1333 and LAZ1366, coupled with SEM fracture morphology. The UTS of as-rolled LAZ1333 alloy and LAZ1366 alloy are 189 MPa and 186 MPa, and the FE are 18.7% and 19%, respectively. The fractures of the as-rolled LAZ1333 alloy and LAZ1366 alloy are mainly composed of large and deep dimples, showing an obvious ductile fracture characteristic. The improvement of UTS of the as-rolled alloy is mainly due to the combined effect of grain refinement strengthening, dislocations strengthening and dispersion strengthening. Compared with the typical brittle fracture of as-solutionized alloys, the transformation of fracture mode of as-rolled alloys is mainly due to the dynamic precipitation of AlLi phase caused by hot rolling, and the precipitated AlLi phase consumes the solid solution at grain boundaries and within grains The reduction in the content of solute atoms in the matrix reduces the lattice distortion, thus reducing the occurrence of intergranular fracture of the alloy. Figure 13 shows the tensile stress–strain curves for as-aged LAZ1333 and LAZ1366, coupled with SEM fracture morphology. The UTS of the as-aged LAZ1333 alloy and LAZ1366 alloy are 184 MPa and 187 MPa, which are 65.8% and 14% higher than those of the as-cast alloy. It is worth noting that the FE of the as-aged LAZ1333 alloy is significantly improved, and its value is 32%. From the SEM fracture morphology, the as-aged LAZ1333 alloy contains many small dimples in addition to large and deep dimples compared with the as-aged LAZ1366 alloy

Figure 12. Room temperature tensile stress–strain curves and fracture morphology of the hot-rolled LAZ13xy (x = 3, 6 wt%, y = 3, 6 wt%) alloys: (**a**) tensile stress–strain curves, fracture morphology of hot-rolled (**b**) LAZ1333 and (**c**) LAZ1366.

Figure 13. Room temperature tensile stress–strain curves and fracture morphology of the as-aged LAZ13xy (x = 3, 6 wt%, y = 3, 6 wt%) alloys: (**a**) tensile stress–strain curves, fracture morphology of as-aged (**b**) LAZ1333 and (**c**) LAZ1366.

The UTS and FE of LAZ13xy (x = 3, 6 wt%, y = 3, 6 wt%) alloys in the different states are shown in Figure 14. In general, the as-aged LAZ1333 alloy has the best mechanical properties, with its UTS increased by 65.8% and its FE increased by 89.3% compared with the as-cast LAZ1333 alloy. The reason why the UTS and FE of the as-aged LAZ1366 alloy cannot be greatly improved is mainly due to the excess precipitation of the AlLi softening phase and the solid solution of excess Zn element.

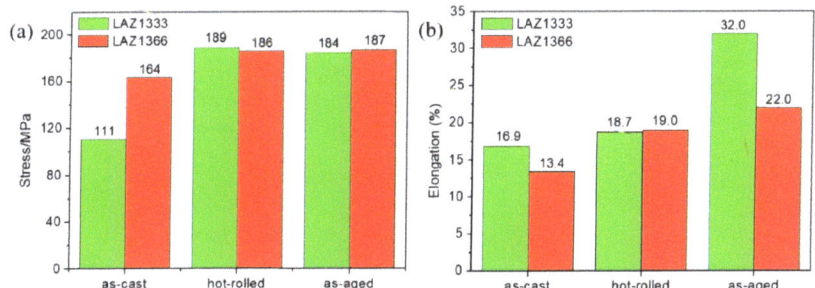

Figure 14. Histograms of mechanical properties of the LAZ13*xy* (*x* = 3, 6 wt%, *y* = 3, 6 wt%) alloys in different states: (**a**) ultimate tensile strength (UTS), (**b**) fracture elongation (FE).

3.3. Damping Properties

Figure 15 shows damping values Q^{-1} of the LAZ13*xy* (*x* = 3, 6 wt%, *y* = 3, 6 wt%) alloys in different states as a function of strain amplitude ε. All curves can be divided into two parts. Under the low-strain amplitude stage (region 1), Q^{-1} is insensitive and grows slowly with ε, which is referred to as the strain-amplitude-independent damping. When the strain amplitude exceeds a critical value ε_{cr} (high-strain amplitude stage), the Q^{-1} substantially increases with ε (region 2), which is referred to as the strain-amplitude-dependent damping [26]. In the as-cast alloy, the LAZ1366 alloy has more solute atoms compared to LAZ1333 alloy. It is well known that the dislocations are weakly pinned by solute atoms and other point defects during the low-strain amplitude stage (region 1), and the dislocation motion between the weak pinning points generates energy dissipation. The longer the distance between the weak pinning points, the more energy is dissipated by the dislocation motion and the higher the damping capacity of the alloy [27]. The greater number of solute atoms shortens the distance of dislocation movement, which is detrimental to the damping capacity of the alloy [28,29]. Therefore, the damping capacity of as-cast LAZ1366 alloy is lower in the low-strain amplitude stage compared to as-cast LAZ1333 alloy. After heat treatments and plastic deformations, the damping capacity of the as-aged alloys are further improved, and their damping value Q^{-1} in the low-strain amplitude stage is larger than 0.014, which is approximately twice that of the as-cast alloys. Compared with the as-cast alloys, a finely dispersed AlLi phase is present in the as-aged alloys. The precipitation of the second phase consumes the solute atoms in the matrix, and the decrease in the number of solute atoms leads to an increase in the distance of dislocation movement, thereby improves the damping capacity of the alloy in the low-strain amplitude region. In the high-strain amplitude stage (region 2), the dislocations are still bound by the strong pinning point such as precipitates, grain boundaries, twin boundaries, stacking faults, etc. The damping value Q^{-1} of the alloy in different states display a sharply increasing trend with an increase of strain amplitude ε. Totally, the damping value Q^{-1} in the high-strain amplitude stage of the as-aged alloys are obviously higher than that of the as-cast alloys. After ageing treatment, the primary phases in the as-cast alloy undergo dissolution and reprecipitation and the distances between AlLi phases increase slightly, thus increases the range of dislocation motion in the matrix. Therefore, the as-aged alloys have the best damping capacity.

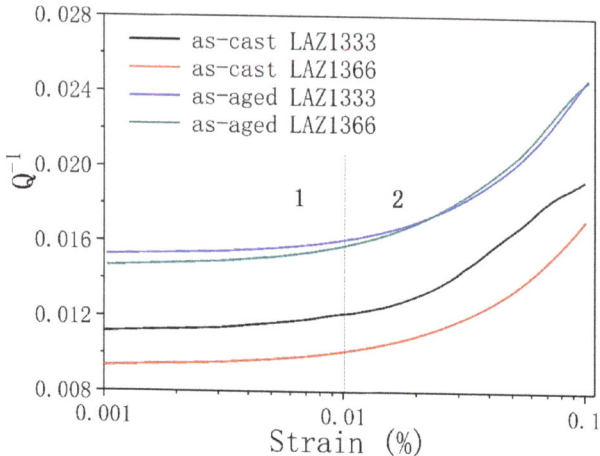

Figure 15. Dependence of the damping capacity of the LAZ13xy (x = 3, 6 wt%, y = 3, 6 wt%) alloys in different states as a function of strain amplitude tested.

4. Conclusions

(1) Both as-cast LAZ1333 alloy and LAZ1366 alloy are composed of β-Li phase and AlLi phase. After the solid solution treatment, the AlLi phase is dissolved, but the hot rolling and subsequent aging make the AlLi phase re-precipitate and disperse in the matrix;

(2) The as-aged LAZ1333 alloy has the best mechanical properties, and its UTS and FE are 184 MPa and 32%, respectively. The UTS and FE of as-aged LAZ1333 alloy increased by 65.8% and 89.3% higher than that of the as-cast alloy. It can be mainly attributed to the combined effect of grain refinement strengthening, dislocations strengthening and dispersion strengthening;

(3) The reason why the UTS and FE of the as-aged LAZ1366 alloy cannot be greatly improved is mainly due to the excess precipitation of AlLi softening phase and the solid solution of excess Zn element;

(4) The as-aged state of LAZ1333 and LAZ1366 alloys exhibits higher damping capacity than as-cast state, and their damping value in the low-strain amplitude stage is larger than 0.014. The improvement in damping capacity is mainly attributed to the increase in the distance of dislocation movement due to the decrease in the number of solute atoms in the matrix.

Author Contributions: Validation, L.H. and S.Y.B.; formal analysis, X.Y. and Y.J.; investigation, X.Y., D.W. and V.S.; data curation, J.W.; writing—original draft preparation, X.Y.; writing—review and editing, X.M. and I.I.T.-B.; supervision, R.W. All authors have read and agreed to the published version of the manuscript.

Funding: This paper was supported by National Natural Science Foundation of China (51871068, 51971071, 52011530025, U21A2049, 52271098), Fundamental Research Funds for the Central Universities (3072022QBZ1002).

Data Availability Statement: The data presented in this study are available in this article.

Conflicts of Interest: The authors declare no conflict of interest.

References

1. Wang, J.; Du, C.; Wu, R.; Xu, L.; Feng, J.; Zhang, J.; Hou, L.; Liu, M.; Liu, B. Effect of Li content on electromagnetic shielding effectiveness in binary Mg–Li alloys: A combined experimental and first-principles study. *J. Mater. Sci. Mater. Electron.* **2022**, *33*, 3891–3900. [CrossRef]
2. Wang, J.; Xu, L.; Wu, R.; Feng, J.; Zhang, J.; Hou, L.; Zhang, M. Enhanced electromagnetic interference shielding in a duplex-phase Mg–9Li–3Al–1Zn alloy processed by accumulative roll bonding. *Acta Metall. Sin. (Engl. Lett.)* **2020**, *33*, 490–499. [CrossRef]
3. Hou, P.; Li, F.; Wu, R.; Gao, R.; Zhang, A. Annealing coordinates the deformation of shear band to improve the microstructure difference and simultaneously promote the strength-plasticity of composite plate. *Mater. Des.* **2022**, *219*, 110696.
4. Ren, L.; Quan, G.; Xu, Y.; Yin, D.; Lu, J.; Dang, J. Effect of heat treatment and predeformation on damping capacity of cast Mg–Y binary alloys. *J. Alloys Compd.* **2017**, *699*, 976–982. [CrossRef]
5. Ma, X.; Jin, S.; Wu, R.; Wang, J.; Wang, G.; Krit, B.; Betsofen, S. Corrosion behavior of Mg–Li alloys: A review. *Trans. Nonferrous Met. Soc. China* **2021**, *31*, 3228–3254. [CrossRef]
6. Yan, H.; Zhou, X.; Gao, X.; Chen, J.; Xia, W.; Su, B.; Song, M. Development of the finegrained Mg–0.6Zr sheets with enhanced damping capacity by high strain rate rolling. *Mater. Charact.* **2021**, *172*, 110826. [CrossRef]
7. Wang, J.; Sun, D.; Wu, R.; Du, C.; Yang, Z.; Zhang, J.; Hou, L. A good balance between mechanical properties and electromagnetic shielding effectiveness in Mg–9Li–3Al–1Zn alloy. *Mater. Charact.* **2022**, *188*, 111888. [CrossRef]
8. Dobkowska, A.; Adamczyk, B.; Kuc, D.; Hadasik, E.; Płociński, T.; Bińczyk, E.; Mizera, J. Influence of bimodal grain size distribution on the corrosion resistance of Mg–4Li–3Al–1Zn (LAZ431). *J. Mater. Res. Technol.* **2021**, *13*, 346–358. [CrossRef]
9. Dobkowska, A.; Adamczyk, B.; Towarek, A.; Maj, P.; Bińczyk, E.; Momeni, M.; Kuc, D.; Hadasik, E.; Mizera, J. The influence of microstructure on corrosion resistance of Mg–3Al–1Zn–15Li (LAZ1531) alloy. *J. Mater. Eng. Perform.* **2020**, *29*, 2679–2686. [CrossRef]
10. Ji, H.; Peng, X.; Zhang, X.; Liu, W.; Wu, G.; Zhang, L.; Ding, W. Balance of mechanical properties of Mg–8Li–3Al–2Zn–0.5Y alloy by solution and low-temperature aging treatment. *J. Alloys Compd.* **2019**, *791*, 655–664. [CrossRef]
11. Sun, Y.; Wang, R.; Peng, C.; Wang, X. Microstructure and corrosion behavior of as-homogenized Mg–xLi–3Al–2Zn0.2Zr alloys (x = 5, 8, 11 wt%). *Mater. Charact.* **2020**, *159*, 110031. [CrossRef]
12. Peng, X.; Liu, W.; Wu, G.; Ji, H.; Ding, W. Plastic deformation and heat treatment of Mg–Li alloys: A review. *J. Mater. Sci. Technol.* **2022**, *99*, 193–206. [CrossRef]
13. Ji, Q.; Wang, Y.; Wu, R.; Wei, Z.; Ma, X.; Zhang, J.; Hou, L.; Zhang, M. High specific strength Mg–Li–Zn–Er alloy processed by multi deformation processes. *Mater. Charact.* **2020**, *160*, 110135. [CrossRef]
14. Sun, Y.; Wang, R.; Ren, J.; Peng, C.; Feng, Y. Hot deformation behavior of Mg–8Li–3Al–2Zn–0.2Zr alloy based on constitutive analysis, dynamic recrystallization kinetics, and processing map. *Mech. Mater.* **2019**, *131*, 158–168. [CrossRef]
15. Liang, X.; Peng, X.; Ji, H.; Liu, W.; Wu, G.; Ding, W. Microstructure and mechanical properties of as-cast and solid solution treated Mg–8Li–xAl–yZn alloys. *Trans. Nonferrous Met. Soc. China* **2021**, *31*, 925–938. [CrossRef]
16. Tang, S.; Xin, T.; Xu, W.; Miskovic, D.; Sha, G.; Quadir, Z.; Ringer, S.; Nomoto, K.; Birbilis, N.; Ferry, M. Precipitation strengthening in an ultralight magnesium alloy. *Nat. Commun.* **2019**, *10*, 1003. [CrossRef] [PubMed]
17. Jin, S.; Ma, X.; Wu, R.; Li, T.; Wang, J.; Krit, B.; Hou, L.; Zhang, J.; Wang, G. Effect of carbonate additive on the microstructure and corrosion resistance of plasma electrolytic oxidation coating on Mg–9Li–3Al alloy. *Int. J. Miner. Metall. Mater.* **2022**, *29*, 1453–1463. [CrossRef]
18. Wang, J.; Wu, R.; Feng, J.; Zhang, J.; Hou, L.; Liu, M. Recent advances of electromagnetic interference shielding Mg matrix materials and their processings: A review. *Trans. Nonferrous Met. Soc. China* **2022**, *32*, 1385–1404. [CrossRef]
19. Zhang, D.; Shi, G.; Zhao, X.; Qi, F. Microstructure evolution and mechanical properties of Mg–x%Zn–1%Mn (x = 4, 5, 6, 7, 8, 9) wrought magnesium alloys. *Trans. Nonferrous Met. Soc. China* **2011**, *21*, 15–25. [CrossRef]
20. Mirzadeh, H. A comparative study on the hot flow stress of Mg–Al–Zn magnesium alloys using a simple physically-based approach. *J. Magnes. Alloy.* **2014**, *2*, 225–229. [CrossRef]
21. Wang, J.; Xu, L.; Wu, R.; An, D.; Wei, Z.; Wang, J.; Feng, J.; Zhang, J.; Hou, L.; Liu, M. Simultaneous achievement of high electromagnetic shielding effectiveness (X-band) and strength in Mg–Li–Zn–Gd/MWCNTs composite. *J. Alloys Compd.* **2021**, *882*, 160524. [CrossRef]
22. Li, W.; Deng, K.; Zhang, X.; Nie, K.; Xu, F. Effect of ultra-slow extrusion speed on the microstructure and mechanical properties of Mg–4Zn–0.5Ca alloy. *Mater. Sci. Eng. A Struct.* **2016**, *677*, 367–375. [CrossRef]
23. Wang, C.; Kang, J.; Deng, K.; Nie, K.; Liang, W.; Li, W. Microstructure and mechanical properties of Mg–4Zn–xGd (x = 0, 0.5, 1, 2) alloys. *J. Magnes. Alloy* **2020**, *8*, 441–451. [CrossRef]
24. Liu, W.; Zeng, Z.; Hou, H.; Zhang, J.; Zhu, Y. Dynamic precipitation behavior and mechanical properties of hot-extruded Mg89Y4Zn2Li5 alloys with different extrusion ratio and speed. *Mater. Sci. Eng. A* **2020**, *798*, 140121. [CrossRef]
25. Tang, S.; Xin, T.; Luo, T.; Ji, F.; Li, C.; Feng, T.; Lan, S. Grain boundary decohesion in body-centered cubic Mg–Li–Al alloys. *J. Alloys Compd.* **2022**, *902*, 163732. [CrossRef]
26. Wang, J.; Jin, Y.; Wu, R.; Wang, D.; Qian, B.; Zhang, J.; Hou, L. Simultaneous improvement of strength and damping capacities of Mg–8Li–6Y–2Zn alloy by heat treatment and hot rolling. *J. Alloys Compd.* **2022**, *927*, 167027. [CrossRef]

27. Wang, D.; Liu, S.; Wu, R.; Zhang, S.; Wang, Y.; Wu, H.; Zhang, J.; Hou, L. Synergistically improved damping, elastic modulus and mechanical properties of rolled Mg–8Li–4Y–2Er–2Zn–0.6Zr alloy with twins and long-period stacking ordered phase. *J. Alloys Compd.* **2021**, *881*, 160663. [CrossRef]
28. Wang, D.; Wu, H.; Wu, R.; Wang, Y.; Zhang, J.; Betsofen, S.; Krit, B.; Hou, L.; Nodir, T. The transformation of LPSO type in Mg–4Y–2Er–2Zn–0.6Zr and its response to the mechanical properties and damping capacities. *J. Magnes. Alloy.* **2020**, *8*, 793–798. [CrossRef]
29. Qian, B.Y.; Wu, R.Z.; Sun, J.F.; Zhang, J.H.; Hou, L.G.; Ma, X.C.; Wang, J.H.; Hu, H.T. Evolutions of Microstructure and Mechanical Properties in Mg–5Li–1Zn–0.5Ag–0.5Zr–xGd Alloy. *Acta Metall. Sin. (Engl. Lett.)* **2023**. [CrossRef]

Disclaimer/Publisher's Note: The statements, opinions and data contained in all publications are solely those of the individual author(s) and contributor(s) and not of MDPI and/or the editor(s). MDPI and/or the editor(s) disclaim responsibility for any injury to people or property resulting from any ideas, methods, instructions or products referred to in the content.

MDPI AG
Grosspeteranlage 5
4052 Basel
Switzerland
Tel.: +41 61 683 77 34

Metals Editorial Office
E-mail: metals@mdpi.com
www.mdpi.com/journal/metals

Disclaimer/Publisher's Note: The statements, opinions and data contained in all publications are solely those of the individual author(s) and contributor(s) and not of MDPI and/or the editor(s). MDPI and/or the editor(s) disclaim responsibility for any injury to people or property resulting from any ideas, methods, instructions or products referred to in the content.